High Temperature Superconductors
and Other Superfluids

High Temperature Superconductors and Other Superfluids

A. S. Alexandrov and Sir Nevill Mott

Routledge
Taylor & Francis Group

LONDON AND NEW YORK

First published 1994 by Taylor & Francis

2 Park Square, Milton Park, Abingdon, Oxon OX14 4RN
711 Third Avenue, New York, NY 10017, USA

Routledge is an imprint of the Taylor & Francis Group, an informa business

First issued in paperback 2017

British Library Cataloguing in Publication Data

A catalogue record for this book is available from the British Library

ISBN 978-0-7484-0309-7 (hbk)
ISBN 978-1-138-07155-1 (pbk)

Contents

Preface

A proper discussion of any theory of superconductivity inevitably involves much advanced mathematics. This preface attempts an expression in simple terms of some of the arguments expressed in this book.

The simplest superfluid to be discovered was superfluid helium He^4. Since the discovery of quantum mechanics it has been known that the helium atom is a boson, and that the atoms of liquid helium should obey the Bose -Einstein statistics. Liquid helium can now be described by the two-fluid model; below the transition temperature T_c it can be considered as a mixture of the superfluid, without entropy, and the normal fluid. If interaction between the particles is neglected, these statistics give

$$k_B T_c = \frac{3.3 \hbar^2 n^{2/3}}{m}$$

where n is the density of particles and m their mass. The transition is of second order, and this formula gives $T_c \simeq 2.8K$ to be compared with $\simeq 2.2K$ observed. The discrepancy is doubtless due to interaction between the particles. It has not yet been possible to calculate it, and it is perhaps surprising that it is so small. It is now known that, also as a result of interaction, the condensate density is only 0.1 of the total one, the remainder being similar to that of a normal liquid. This and also the weakly interacting Bose-gas are discussed in Chapter 1.

The BCS theory of superconductivity, highly successful for metals and alloys with $T_c < 20K$, is based on the demonstration by Fröhlich that electrons in states near the Fermi surface of a metal can attract each other weakly on account of their interaction with phonons. Cooper showed that pairs (bosons) could thereby be formed, of diameter much greater than the lattice parameter, and stable only through their interaction with the carriers of the Fermi sea. The final theory showed that in a small volume of momentum space round the Fermi surface, the electrons would form pairs, bosons, all having the same energy. They would leave an energy gap of width given by

$$\Delta(0) \simeq 2\hbar\omega_D exp(-1/\lambda)$$

where $\hbar\omega_D/k_B$ is the Debye temperature, $\lambda = VN(0)$, V represents the electron-electron coupling due to the electron-phonon interaction, and $N(0)$ is the density of states at the Fermi level. λ is normally about $0.1 - 0.3$. As the temperature is raised, electrons are excited across the gap, giving a second order phase transition at T_c where

$$k_B T_c \simeq 1.14\hbar\omega_D exp(-1/\lambda)$$

We discuss the BCS theory and its canonical extension to intermediate coupling in Chapter 2 and Chapter 3 respectively.

In the new superconductors T_c approaches the Debye temperature, and the natural approach is to ask what happens if λ approaches unity or more when the correlation length (the radius of the pairs) decreases and the gap widens. However, the insight which leads to the theory of this book was that $\lambda = 1$ is also the criterion for the formation of a polaron by each electron. A rapid and possibly discontinuous transition takes place in which each electron forms a polaron, the pairs form bipolarons which are heavy, so that we are left with a degenerate gas of bipolarons. A full mathematical discussion of this is given in Chapters 4 and 5. This can only occur if the density of electrons is not too large, as to form a polaron involves several surrounding ions. A criticism has been made that such bipolarons would be too heavy to move, but we do not believe this to be the case. The critical temperature will then be given by the free boson formula; m we believe to be about $10m_e$, and n should be the number of electron (hole) pairs not localised by disorder. This and some alternative theories of high-Tc oxides are discussed in Chapter 6.

From a theoretical point of view the simplest of the new superconductors should be cubic bismuth oxides, but because of experimental difficulties in specimen preparation we do not have firm evidence that they are of this class. In our view the strongest evidence would be an examination of the heat conductivity. The large contribution from the phonons has to be subtracted. The term K due to the current carriers should obey the Wiedemann-Franz ratio

$$\frac{K}{T\sigma} = 3 \left(\frac{k_B}{e^*}\right)^2$$

where e^* is the charge on the carrier and σ the electrical conductivity. In Chapter 7 we discuss the evidence that $e^* = 2e$ for the copper oxide materials. These have been the subject of the great majority of experimental work. They are all of the form of doped Mott insulators. Thus $La_{2-x}Sr_xCuO_4$ one of the first to be investigated is for $x = 0$ an antiferromagnetic insulator, strongly anisotropic, with a band gap of about $2eV$ and Néel temperature c.$200K$. Doping with Sr (or addition of oxygen) introduces holes. The holes are thought to be in the oxygen $2p$ band, hybridised with $Cu3d9$. A small doping destroys the antiferromagnetism, leading to a 'spin glass' state in which holes, which have already formed pairs, move by activated variable-range hopping. At x=0.02 there is a transition to a metallic state, which (because the current carriers are pairs) is a superconductor. The number of carriers is at first small, so there should be no difficulty in finding anions with which to form polarons. As x increases, n increases and so does T_c. Eventually T_c saturates, and this we believe is because of overcrowding: the carriers are too many for the available ions.

The problem of what happens when a Mott insulator is doped is often stated in the literature to be unsolved. However, it is known that in doped magnetic *semiconductors* the carriers form spin polarons (see Chapter 4). These are clusters of magnetic moments oriented in the same direction opposite to the moment on the carriers. We assume in this book that the same happens in a Mott insulator in which the moment is $S = 1/2$. Our carriers are then rather complicated; in the centre we believe a spin polaron forms, with round it, on account of the high static dielectric constant, a polarised region. As a consequence the effective mass of the bipolaron will depend on phonon frequency and hence on oxygen, which explains the isotope effect .

The greater part of this book is devoted to applying the theory to explain the behavior of the copper oxide superconductors both above and below the transition point T_c. Phenomena treated include the nuclear magnetic resonance and neutron scattering, resistivity and Hall 'constant' above T_c, the thermopower and infrared conductivity, the isotope effect and heat capacity, the heat transport, critical magnetic fields, the results obtained from angle resolved photoemission, and some other observations (Chapter 7). We claim that no effect known to us is qualitatively in disagreement with the theory. Of course new effects may be discovered which are. Meanwhile we hope that the book will be a challenge to look for that.

Chapter 1

Liquid helium and Bose-gas

1.1 Liquid helium and Bose-Einstein condensation

This book is about a certain class of superfluids, namely the high temperature superconductors. Perhaps the simplest of the superfluids is liquid He^4. Since the very early days of quantum mechanics it has been known that the helium atom He^4 is a boson; it is built of an even number of particles (two electrons, two neutrons and two protons), so that a wave function $\psi(\mathbf{r}_1, \mathbf{r}_2)$ must be symmetrical in $\mathbf{r}_1 - \mathbf{r}_2$ where $\mathbf{r}_1, \mathbf{r}_2$ are the electron's coordinates. This has the result that half the rotational states of the molecule He_2 are absent and that the scattering of alpha particles by helium for low energies deviates from the Rutherford scattering law. It also has the remarkable result that this is not so for the light isotope He^3, which shows that the number of neutrons in the nucleus affects properties depending otherwise only on the electrons and nuclear charge.

It is now recognised that the properties of the superfluid depend on the Bose-Einstein statistics obeyed by the particles in liquid helium. Bose statistics of non-interacting particles means that the number of particles in each one-particle quantum state ν is given

1

by

$$n_\nu = \frac{1}{exp(\frac{\epsilon_\nu - \mu}{k_B T}) - 1} \tag{1.1}$$

where ϵ_ν is the one-particle energy, μ is the chemical potential and T the temperature. At low temperatures the major part of noninteracting bosons is in the ground state with $\epsilon_\nu = 0$. The chemical potential is the difference of the ground state energies of the system with $N + 1$ and N particles, $\mu = E_0(N + 1) - E_0(N)$, and it is zero in this case. The distribution function has to satisfy the condition

$$\sum_\nu n_\nu = n \tag{1.2}$$

In a macroscopic system the one particle energy levels are very close to each other so the density of states (DOS) $N(\epsilon)$ is a continuous function

$$N(\epsilon) = \sum_\nu \delta(\epsilon - \epsilon_\nu). \tag{1.3}$$

In the absence of an external field the one particle eigenstates are classified with the wave vector $\mathbf{k}$ and $\epsilon_\nu = \hbar^2 k^2/2m$. In this case for three dimensions

$$N(\epsilon) = \frac{m^{3/2}}{\sqrt{2}\pi^2\hbar^3}\sqrt{\epsilon} \tag{1.4}$$

We take the volume of the system $\Omega = 1$. With an exception of the lowest energy state with $\epsilon_\nu = 0$ the distribution function has no singularities and the sum over all excited states in Eq.(1.2) can be replaced for the integral with a smooth function $N(\epsilon)$. Therefore the number of particles in the lowest energy state n_0 is determined by

$$n_0 = n - \int_{0+}^{\infty} \frac{d\epsilon N(\epsilon)}{exp(\epsilon/k_B T) - 1} \tag{1.5}$$

For free bosons the integral in Eq.(1.5) is $n(T/T_c)^{3/2}$ and

$$n_0 = n\left(1 - (T/T_c)^{3/2}\right) \tag{1.6}$$

where

$$k_B T_c = \frac{\hbar^2(\sqrt{2}\pi^2 n/\Gamma(3/2)\zeta(3/2))^{2/3}}{m} \tag{1.7}$$

is the critical temperature of the Bose-Einstein condensation. With $\Gamma(3/2) = \sqrt{\pi}/2$ and $\zeta(3/2) \simeq 2.612$

$$k_B T_c \simeq \frac{3.31\hbar^2 n^{2/3}}{m} \tag{1.8}$$

Below T_c a macroscopic fraction of particles n_0 are in the lowest one-particle state. The transition at $T = T_c$ in the ideal Bose-gas is a phase transition of the 'third' kind, driving by the quantum mechanical kinematic interaction. The heat capacity has a kink at $T = T_c$.

With the non-interacting model, Eq.(1.8) one obtains for He^4 $T_c \simeq 2.8K$ which is surprisingly close to the experimental value $2.17K$ despite of the fact that the dynamical interaction is strong in liquid helium. This interaction is responsible for the heat capacity anomaly at T_c, which in He^4 has a well known λ-like divergent shape, Fig.1.1.

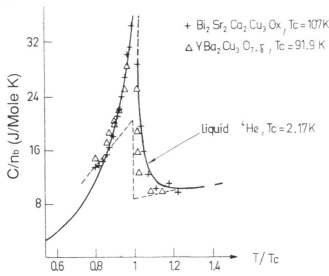

Fig.1.1. Heat capacity anomaly in two high-T_c oxides compared with He^4 ($n_b = 1$), solid line and with the BCS curve, dashed line.

Superfluidity itself is due to the dynamical interaction. To see this one can apply the Landau criterion of superfluidity. Let us consider a liquid flowing at a constant velocity $\mathbf{v}$. We assume following Landau that the excited states of the liquid can be classified with an elementary excitation energy $\epsilon(\mathbf{p})$, the excitation occupation numbers $n(\mathbf{p})$ and with the momentum $\mathbf{p}$ due to space homogeneity. Because of the friction the kinetic energy of the liquid would be dissipated and the flow would gradually become slower. If we take a coordinate frame moving with the liquid this process would correspond to the appearance at least one elementary excitation in the liquid.

The momentum and energy conservation yields

$$-M\mathbf{v} = M\mathbf{v}' + \mathbf{p} \tag{1.9}$$

and

$$\frac{Mv^2}{2} = \frac{Mv'^2}{2} + \epsilon(\mathbf{p}) \tag{1.10}$$

where M is a mass of a classical object (such as a retaining wall) exciting the liquid, v, v' are velocities of the object before and after collision with the liquid. Combining Eq.(1.9) and Eq.(1.10) one obtains

$$\epsilon(\mathbf{p}) + \mathbf{p} \cdot \mathbf{v} + \frac{p^2}{2M} = 0 \tag{1.11}$$

A minimum value of the velocity v_c which satisfies to this equation is

$$v_c = min\frac{\epsilon(\mathbf{p})}{p} \tag{1.12}$$

The liquid is superfluid if $v_c \neq 0$ and $v < v_c$ when it cannot be excited. For the ideal gas $\epsilon(\mathbf{p}) = p^2/2m$ and consequently $v_c = 0$. Therefore the Bose condensate alone does not lead to superfluidity. A non-zero critical velocity is a result of the dynamical interaction of excitations with the condensate as follows from the Bogoliubov (1947) theory of superfluidity of weakly interacting bosons.

1.2   Weakly interacting Bose-gas

If the interaction is weak one can expect that the occupation numbers of one-particle states are not very much different from those in the ideal Bose-gas. In particular the state with zero momentum $k = 0$ remains to be macroscopically occupied and the corresponding Fourier component of the field operator $\psi(\mathbf{r})$ has anomalously large matrix element between the ground states of the system containing $N + 1$ and N bosons. It is convenient to open the system, introducing a chemical potential μ. In this case the quantum state is a superposition of states $|N>$ with slightly different total numbers of bosons. The weight of each state is a smooth function of N which is practically constant near the average number $\bar{N}$ on the scale $\pm\sqrt{N}$. Because ψ changes the number of particles only by one its *diagonal* matrix element coincides with the off-diagonal, calculated for the states with fixed $N = \bar{N}+1$ and $N = \bar{N}$. Following Bogoliubov (1947) one can separate the large diagonal matrix element ψ_0 from ψ treating the remainder $\tilde{\psi}$ as a small fluctuation

$$\psi(\mathbf{r}) = \psi_0 + \tilde{\psi}(\mathbf{r}). \qquad (1.13)$$

The *anomalous* average $\psi_0 = \langle\psi(\mathbf{r})\rangle$ is equal to $\sqrt{n_0}$ in a homogenuos system.

Substituting Eq.(1.13) into the hamiltonian of the interacting Bose-gas

$$
\begin{aligned}
H &= -\int d\mathbf{r}\psi^\dagger(\mathbf{r})\left(\frac{\nabla^2}{2m} + \mu\right)\psi(\mathbf{r}) \\
&+ \frac{1}{2}\int d\mathbf{r}d\mathbf{r}'V(\mathbf{r}-\mathbf{r}')\psi^\dagger(\mathbf{r})\psi^\dagger(\mathbf{r}')\psi(\mathbf{r})\psi(\mathbf{r}') \qquad (1.14)
\end{aligned}
$$

and restricting oneself by terms of the second order in $\tilde{\psi}$ one arrives at the reduced hamiltonian

$$H = -\mu n_0 - \int d\mathbf{r}\tilde{\psi}^\dagger(\mathbf{r})\left(\frac{\nabla^2}{2m} + \mu\right)\tilde{\psi}(\mathbf{r}) + H_{int} \qquad (1.15)$$

where

$$H_{int} = \frac{n_0}{2}\int d\mathbf{r}d\mathbf{r}'V(\mathbf{r}-\mathbf{r}')\left(\tilde{\psi}^\dagger(\mathbf{r})\tilde{\psi}^\dagger(\mathbf{r}') + \tilde{\psi}(\mathbf{r})\tilde{\psi}(\mathbf{r}') + 2\tilde{\psi}^\dagger(\mathbf{r})\tilde{\psi}(\mathbf{r}')\right).$$

For an open system $E - \mu\bar{N}$ has minima for the eigenstates. That is why the one-particle energies in Eq.(1.14) and Eq.(1.15) are related to μ. Here and further $\hbar = c = k_B = 1$ unless specified and the averaged interaction potential is assumed to be zero

$$\int d\mathbf{r}V(\mathbf{r}) = 0 \qquad (1.16)$$

which is normally the case for charged particles with a uniform background of the opposite charge. If it is not the case a zero Fourier component of the interaction, Eq.(1.16) can be included in the chemical potential μ.

The reduced hamiltonian Eq.(1.15) being quadratic in the operators can be *exactly* diagonalised with the linear transformation of the field operator to the k-space quasiparticle operator $\alpha_{\mathbf{k}}$

$$\tilde{\psi}(\mathbf{r}) = \sum_{\mathbf{k}}(u_{\mathbf{k}}\alpha_{\mathbf{k}} + v_{\mathbf{k}}\alpha^{\dagger}_{-\mathbf{k}})e^{i\mathbf{k}\cdot\mathbf{r}} \qquad (1.17)$$

with the coefficients

$$u_{\mathbf{k}}^2 = \frac{1}{2}\left(1 + \frac{\xi_{\mathbf{k}}}{\epsilon_{\mathbf{k}}}\right) \qquad (1.18)$$

$$v_{\mathbf{k}}^2 = -\frac{1}{2}\left(1 - \frac{\xi_{\mathbf{k}}}{\epsilon_{\mathbf{k}}}\right) \qquad (1.19)$$

and

$$u_{\mathbf{k}}v_{\mathbf{k}} = -\frac{V(\mathbf{k})n_0}{2\epsilon_{\mathbf{k}}} \qquad (1.20)$$

where $\xi_{\mathbf{k}} = \frac{k^2}{2m} - \mu + V(\mathbf{k})n_0$ and $V(\mathbf{k})$ the Fourier component of the interaction potential. Choosing the coefficients in the form Eq.(1.18-20) one eliminates the off-diagonal terms proportional to $\alpha^{\dagger}\alpha^{\dagger}$ and $\alpha\alpha$ in the hamiltonian and retains the bosonic statistics for quasiparticles: $\alpha_{\mathbf{k}}\alpha^{\dagger}_{\mathbf{k}'} - \alpha^{\dagger}_{\mathbf{k}'}\alpha_{\mathbf{k}} = \delta_{\mathbf{k},\mathbf{k}'}$. The transformed hamiltonian describes the ideal Bose-gas of *quasiparticles* with a new energy dispersion $\epsilon_{\mathbf{k}}$:

$$H = E_0 + \sum_{\mathbf{k}}\epsilon_{\mathbf{k}}\alpha^{\dagger}_{\mathbf{k}}\alpha_{\mathbf{k}} \qquad (1.21)$$

where E_0 is the ground state energy:

$$E_0 = -\mu n_0 + \frac{1}{2}\sum_{\mathbf{k}}(\epsilon_{\mathbf{k}} - \xi_{\mathbf{k}}) \qquad (1.22)$$

and

$$\epsilon_{\mathbf{k}} = \sqrt{\left(\frac{k^2}{2m} - \mu\right)^2 + 2V(\mathbf{k})n_0\left(\frac{k^2}{2m} - \mu\right)} \qquad (1.23)$$

The number of quasiparticles as well as the number of condensed bosons n_0 are not fixed. The first one is determined by the Bose-Einstein distribution function $f_{\mathbf{k}} = (exp\epsilon_{\mathbf{k}}/T - 1)^{-1}$ with zero chemical potential while the second one should conserve the total number of bosons:

$$n_0 = n - \tilde{n}, \qquad (1.24)$$

and minimise the total energy E:

$$\frac{\partial E}{\partial n_0} = 0 \qquad (1.25)$$

Here $E = E_0 + \sum_{\mathbf{k}} \epsilon_{\mathbf{k}} f_{\mathbf{k}}$ and the number of bosons above the condensate $\tilde{n}$ is

$$\tilde{n} = \sum_{\mathbf{k}} u_{\mathbf{k}}^2 f_{\mathbf{k}} + v_{\mathbf{k}}^2(1 + f_{\mathbf{k}}). \qquad (1.26)$$

It is easy to show using Eq.(1.25) that the chemical potential μ is of the same order as the interaction between bosons above the condensate neglected with the hamiltonian Eq.(1.15). Therefore within the Bogoliubov approach $\mu = 0$. As a result the elementary excitation spectrum has a form

$$\epsilon_{\mathbf{k}} = \sqrt{\frac{k^4}{4m^2} + \frac{k^2 V(k) n_0}{m}} \qquad (1.27)$$

It satisfies the Landau criterium $v_c \neq 0$ for any realistic repulsive interaction $V > 0$ including the Coulomb one.

Not all bosons are condensed in the interacting Bose-gas even at $T = 0$. As one can derive from Eq.(1.26) there is a small fraction of them $\tilde{n} \ll n$ above the condensate. The ratio $\tilde{n}/n$ should be small to apply the Bogoliubov approach, which restricts its region by low temperatures and weak interaction. When the interaction becomes large the result is that only a small fraction of bosons in the $\mathbf{k} = 0$ state. This has been shown theoretically and experimentally for He^4 by calculation and neutron diffraction (for more details see Pines (1962) and Svensson (1984)). The remaining particles are still part of the superfluid at $T = 0$, having zero entropy, but their wave functions and distribution are similar to that of a normal fluid. Mott (1993a) has suggested that the energy of the system will be of the order

$$V n_0^2 + \frac{1 - n_0}{ma^2} \qquad (1.28)$$

where V is the short range repulsion (for neutral particles) and a the interparticle distance. The second term is the kinetic energy of particles which are not in the condensate. Minimising we find $n_0 = ma^2/2V$; n_0 will drop with increasing V but not vanish.

The Bogoliubov formula Eq.(1.27) explains qualitatively why the non-interacting model gives fairly satisfactory result for T_c even for a strongly correlated liquid like He^4. If one extrapolates the Bogoliubov spectrum Eq.(1.27) to finite temperatures with a *temperature* dependent $n_0(T)$, then at $T = T_c$ where $n_0(T_c) = 0$ one obtains the spectrum of non-interacting particles. Therefore one can expect that the corrections to the expression Eq.(1.7) due to the interaction come from those to the effective mass of a boson in the normal state, which normally are not very large.

1.3   Charged Bose-gas

The Fourier component of the Coulomb interaction is $V(k) = 4\pi e^{*2}/\epsilon_0 k^2$ where ϵ_0 a dielectric constant of the background and bosonic charge is e^*. This makes (Foldy (1961)):

$$\epsilon_k = \sqrt{\frac{k^4}{4m^2} + \omega_{p0}^2} \qquad (1.29)$$

with a gap $\omega_{p0} = \sqrt{4\pi e^{*2}n_0/\epsilon_0 m}$ which is the classical plasma frequency for a plasma of density n_0, Fig.1.2. The number of bosons pushed up from the condensate by the Coulomb repulsion at $T = 0$ is

$$\tilde{n} = \langle \tilde{\psi}^\dagger(\mathbf{r})\tilde{\psi}(\mathbf{r}) \rangle = \sum_{\mathbf{k}} v_{\mathbf{k}}^2 \qquad (1.30)$$

which is small

$$\tilde{n} \simeq 0.2nr_s^{3/4} \qquad (1.31)$$

if the density of bosons is *high*

$$r_s = \frac{me^{*2}}{\epsilon_0(4\pi n/3)^{1/3}} << 1 \qquad (1.32)$$

This is contrary to the case of hard-core bosons $V(k) = const$, when the weak-interaction condition $\tilde{n} << n$ is satisfied in the *dilute* limit $nr_0^3 << 1$ where r_0 is the core radius. The extension of the spectrum Eq.(1.29) to finite temperatures with $n_0(T)$ yields the temperature dependent gap disappearing at $T = T_c$.

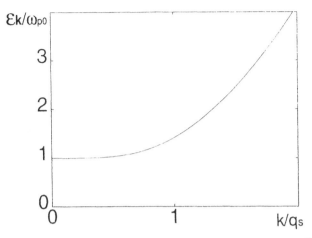

Fig.1.2. Excitation spectrum of a 3d charged Bose-gas, $q_s = \sqrt{2m\omega_{p0}}$.

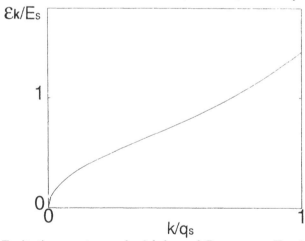

Fig.1.3. Excitation spectrum of a 2d charged Bose-gas at $T = 0$.

1.4 Expulsion of a magnetic field by charged bosons

Any theory of superconductivity should explain the expulsion of a weak magnetic flux from a superconductor (the Meissner-Ochsenfeld effect). For a charged Bose-gas this effect was established by Shafroth (1955). The hamiltonian in an external magnetic field with the vector potential $\mathbf{A}(\mathbf{r})$ is obtained by the replacement in Eq.(1.14) ∇ in the kinetic energy for $\nabla - ie^*\mathbf{A}(\mathbf{r})$. One can choose $\mathbf{A}$ in such a form that $\nabla \cdot \mathbf{A} = 0$ and apply the Bogoliubov displacement transformation Eq.(1.13) with the $\mathbf{r}$ -dependent condensate wave-function $\psi_0(\mathbf{r})$:

$$\psi(\mathbf{r} = \psi_0(\mathbf{r}) + \tilde{\psi}(\mathbf{r}) \qquad (1.33)$$

Minimising the energy with respect to the classical parameter $\psi_0(\mathbf{r})$ and neglecting the fluctuations $\tilde{\psi}$ we arrive at

$$\left(\frac{(\nabla - ie^*\mathbf{A}(\mathbf{r}))^2}{2m} - \int d\mathbf{r}' V(\mathbf{r} - \mathbf{r}')|\psi_0(\mathbf{r}')|^2 \right) \psi_0(\mathbf{r}) = 0 \quad (1.34)$$

The distribution of the magnetic field $\mathbf{H} = curl\mathbf{A}$ in a sample is determined by that of the current density $\mathbf{j}(\mathbf{r})$

$$curlcurl\mathbf{A}(\mathbf{r}) = 4\pi\mathbf{j}(\mathbf{r}) \qquad (1.35)$$

where

$$\mathbf{j}(\mathbf{r}) = \frac{ie^*}{2m} (\psi_0^*(\mathbf{r})\nabla\psi_0(\mathbf{r}) - \psi_0(\mathbf{r})\nabla\psi_0^*(\mathbf{r})) - \frac{e^{*2}\mathbf{A}(\mathbf{r})|\psi_0(\mathbf{r})|^2}{m}$$
$$(1.36)$$

It follows from Eq.(1.34) that in the chosen gauge the linear correction to the homogeneous condensate wave-function is absent. Hence within the linear approximation in the external field

$$\psi_0(\mathbf{r}) = \sqrt{n_0} + O(A^2). \qquad (1.37)$$

This *stiffness* of the condensate wave function with respect to the magnetic field results in the London (1935) equation

$$curlcurl\mathbf{A}(\mathbf{r}) + \lambda_H^{-2}\mathbf{A}(\mathbf{r}) = 0 \qquad (1.38)$$

where

$$\lambda_H = \sqrt{\frac{m}{4\pi e^{*2}n_0}} \qquad (1.39)$$

is the London penetration depth. The weak magnetic field penetrates to the charged Bose-gas only into a skin of a depth λ_H, as one can see solving the London equation Eq.(1.38) for a simple cylindrical geometry.

1.5 Near two-dimensional bosons

For applications it is important to know how the critical temperature and the excitation spectrum are modified with the lowering of
the dimensionality of bosons. We consider a model of free bosons
on a lattice consisting of planes coupled with each other by the
inter-plane hopping integral $t_\perp$. With a three-dimensional correction to the band dispersion taken into account:

$$E_{\mathbf{k}} = \frac{k_\parallel^2}{2m} + 2t_\perp \left(1 - cos(k_\perp d)\right) \tag{1.40}$$

the density of states $N(\epsilon)$ is given by:

$$N(\epsilon) = \frac{1}{2\pi t} arccos(1 - \frac{\epsilon}{2t_\perp}) \tag{1.41}$$

for $0 < \epsilon < 4t_\perp$,

$$N(\epsilon) = 1/2t \tag{1.42}$$

for $4t_\perp < \epsilon < 2t$ and

$$N(\epsilon) = 1 - \frac{1}{2\pi t} arccos(1 - \frac{\epsilon - 2t}{2t_\perp}) \tag{1.43}$$

for $2t < \epsilon < 2t + 4t_\perp$ where $t \gg t_\perp$ the in-plane hopping integral,
$t = \pi/ma^2$; a and d the in-plane and inter-plane lattice constants,
and m the in-plane effective mass.

Substitution of $N(\epsilon)$ into Eq.(1.5) with an assumption $T_c \ll 2t$
yields for $n_0 = 0$:

$$\frac{T_c}{2\pi t} \int_0^{4t_\perp/T_c} dx \frac{arccos(1 - xT_c/2t_\perp)}{exp(x) - 1} - \frac{T_c}{2t} ln \left(1 - exp(-\frac{4t_\perp}{T_c})\right) = n. \tag{1.44}$$

Here n is the number of bosons per cell. This equation is simplified
if $T_c \gg t_\perp$:

$$T_c = \frac{2tn}{L} \tag{1.45}$$

with

$$L = 1 + ln \left(\frac{nt}{t_\perp}\right) \tag{1.46}$$

As a result the critical temperature of the Bose-Einstein condensation decreases with the increasing distance between planes. Because the inter-plane hopping depends on the distance exponentially, T_c drops approximately as d^{-1}. In the limit $d = \infty$ the density of states is constant and the integral in Eq.(1.5) diverges logarithmically, so there is no Bose-Einstein condensation ($T_c = 0$) in this limit. But this is not so when interaction between bosons is taken into account. According to Popov (1976) a two-dimensional hard-core Bose-gas is superfluid below

$$T_c \simeq \frac{2tn}{\ln\ln(a^2/r_0^2 n)}. \tag{1.47}$$

There is no Bose condensate in one or two-dimensions. However bosons with small wave-numbers play its role creating the superfluid density of order of the total one.

In general case of a single particle d spersion $E_{\mathbf{k}}$ the Bogoliubov spectrum of a condensed Bose-gas is $_{\mathbf{g}}$iven by:

$$\epsilon_{\mathbf{k}} = \sqrt{E_{\mathbf{k}}^2 + 2E_{\mathbf{k}}V(\mathbf{k})n_0} \tag{1.48}$$

For charged bosons in the long-wave limit $k_{\|} = 0, k_{\perp} \to 0$ $V(\mathbf{k})$ is the three-dimensional Coulomb interaction $V(\mathbf{k}) \sim 1/k^2$ and the excitation spectrum Eq.(1.48) has a plasma -like temperature dependent gap

$$\omega_{p0} = \sqrt{\frac{4\pi n_0(T)e^{*2}}{\epsilon_0 m_c}} \tag{1.49}$$

with $m_c = 1/2t_{\perp}d^2$ the effective mass in c-direction. The gap Eq.(1.49) depends on temperature, disappearing above T_c because only condensed bosons contribute to it. In a pure two-dimensional system $V(\mathbf{k}) = 2\pi e^{*2}/\epsilon_0 k_{\|}$ and the Bogoliubov spectrum is gapless, Fig. 1.3:

$$\epsilon_{\mathbf{k}} = E_s\sqrt{k_{\|}/q_s + k_{\|}^4/q_s^4} \tag{1.50}$$

with $E_s = q_s^2/2m$, $q_s = (8\pi e^{*2}n_0 m/\epsilon_0)^{1/3}$ a two dimensional screening wave-number and n_0 the in-plane condensate density ($T = 0$).

Chapter 2

BCS theory of superconductivity

The crucial demonstration that superfluidity was linked to Bose particles and Bose-Einstein condensation came after the discovery of He^3 in 1949, which failed to show the characteristic λ transition within a reasonable wide temperature interval around the critical temperature for the onset of superfluidity in He^4. The idea of the bosonization of electrons due to their pairing, introduced by Ogg (1946) became very attractive as an explanation for superconductivity after Schafroth (1955) showed that a gas of charged bosons with charge 2e would indeed show a Meissner effect, which is one of the most crucial features of superconductors (see Chapter 1). However this phenomenological Bose-gas picture was condemned and later on practically forgotten because it was unable to account quantitatively for the critical parameters of the classical (low-temperature) superconductors. Using for T_c the Bose-Einstein condensation temperature Eq.(1.8) one obtains the utterly meaningless result $T_c = 10000K$ with the atomic density of electron pairs $\simeq 10^{22} cm^{-3}$ and with the effective mass $m^{**} = 2m_e$. The failure of the Schafroth picture stems from a very large size ξ of pairs in classical superconductors, which can be estimated from the uncertainty principle:

$$\xi \simeq \frac{1}{\delta p} \tag{2.1}$$

13

The uncertainty of the momentum δp is estimated from the uncertainty of the kinetic energy $\delta E \simeq v_F \delta p$, which should be of the order of T_c, the only characteristic energy of the superconducting state. Therefore

$$\xi \simeq \frac{v_F}{T_c} \qquad (2.2)$$

and turns out to be more than one micron for simple metals with $T_c \simeq 1K$ and the Fermi velocity $v_F \simeq 10^8 cm/s$. That means that pairs in classical superconductors overlap strongly and the Shafroth model of real space pairs is wrong. The finally successful theory for classical superconductors was formulated by Bardeen, Cooper and Schrieffer (1957). In this chapter we present the main points of the BCS theory, using the Bogoliubov transformation (1947,1958) for the weak-coupling regime and the Green's function technique for the intermediate coupling in Chapter 3. More details will be found in the excellent books by Schrieffer(1965), De Gennes (1989) and Abrikosov, Gor'kov and Dzyaloshinskii (1963).

2.1 Ground state and excitations

2.1.1 Fröhlich hamiltonian

In their formulation of the weak-coupling theory of superconductivity BCS used the idea of Fröhlich (1950) that the electron-phonon interaction is responsible. The key point is the fact discovered by Cooper (1957) that any attraction between *degenerate* electrons leads to their pairing; it does not matter how weak it is. One can understand this by transforming the corresponding Schrödinger equation for a pair into momentum space. Because of the Pauli principle a pair of electrons can move only along the Fermi-surface, and it is well known that in two dimensions a bound state exists for any attraction however weak. The Cooper solution of a two-particle problem in the presence of the Fermi sea demonstrates the instability of the Fermi-liquid towards the pair formation and their Bose-Einstein condensation.

To obtain a many-particle superconducting ground state one can start with the Fröhlich hamiltonian, describing the Bloch bands of electrons in a periodic lattice potential and their interaction with the lattice vibrations:

$$H_e = \sum_{i=1}^{N_e} \left(-\frac{\nabla_i^2}{2m_e} + V(\mathbf{r}_i) \right) \tag{2.3}$$

where N_e is the total number of electrons in a crystal of volume Ω and

$$V(\mathbf{r}) = \sum_l v(\mathbf{r} - \mathbf{R}_l) \tag{2.4}$$

is the potential energy in a crystal field distorted by phonons. The interaction of an electron with a single site is described in a Hartree approximation by the potential $v(\mathbf{r})$. In many solids the distance of ions $\mathbf{u}_l = \mathbf{R}_l - l$ from the equilibrium positions l is small compared with a lattice constant a, which allows us to expand $v(\mathbf{r} - \mathbf{R}_l)$ near equilibrium:

$$v(\mathbf{r} - \mathbf{R}_l) \simeq v(\mathbf{r} - l) - \mathbf{u}_l \cdot \nabla v(\mathbf{r} - l) \tag{2.5}$$

A zero order hamiltonian H_0, which includes the electron and ion kinetic energies, the periodic part of the crystal field and the ion-ion interaction can be diagonalized with the Bloch one-particle eigenstates $\psi_{\mathbf{k},n,s}(\mathbf{r})$ and with harmonic phonons by substitution in Eq.(2.5):

$$\mathbf{u}_l = \sum_{\mathbf{q},\nu} \frac{\mathbf{e}_{\mathbf{q},\nu}}{\sqrt{2NM\omega_{\mathbf{q},\nu}}} d_{\mathbf{q},\nu} exp(i\mathbf{q}l) + h.c. \tag{2.6}$$

where $\mathbf{k}, \mathbf{q}$ are electron and phonon momenta; n, ν label electron bands and phonon branches correspondingly, s is a spin, and $d_{\mathbf{q},\nu}$ is a phonon (Bose) annihilation operator with $\mathbf{e}_{\mathbf{q},\nu}$ being a unit vector, $\omega_{\mathbf{q},\nu}$ a phonon frequency, M ionic mass, and N is the number of ions (sites) in a crystal. As a result the total hamiltonian of a crystal, including the electron-phonon interaction H_{e-ph}, in a second quantization has a form:

$$H = H_0 + H_{e-ph}, \tag{2.7}$$

where

$$H_0 = \sum_{\mathbf{k},n,s} \xi_{\mathbf{k},n,s} c^\dagger_{\mathbf{k},n,s} c_{\mathbf{k},n,s} + \sum_{\mathbf{q},\nu} \omega_{\mathbf{q},\nu} d^\dagger_{\mathbf{q},\nu} d_{\mathbf{q},\nu} \tag{2.8}$$

describes independent Bloch electrons and phonons with $c_{\mathbf{k},n,s}$ an electron (fermion) annihilation operator, $\xi_{\mathbf{k},n,s} = E_{\mathbf{k},n,s} - \mu$ is a band energy dispersion. Because the electron-phonon interaction leads to the pairing of electrons it is convenient to consider an open system with a fixed chemical potential μ rather than to fix N_e to avoid some artificial difference between odd and even N_e. For an open system $H - \mu N_e$ has a minimum in a ground state. That is why the electron energy in Eq.(2.8) is related to μ.

The electron-phonon interaction is linear in phonon operators and quadratic in electron ones. We write it as

$$H_{e-ph} = \frac{1}{\sqrt{2N}} \sum_{\mathbf{k},\mathbf{q},n,n',\nu,s} \gamma_{n,n'}(\mathbf{q},\mathbf{k},\nu)\omega_{\mathbf{q},\nu}c^{\dagger}_{\mathbf{k},n,s}c_{\mathbf{k}-\mathbf{q},n',s}d_{\mathbf{q},\nu} + h.c.$$

(2.9)

with a dimensionless matrix element

$$\gamma_{n,n'}(\mathbf{q},\mathbf{k},\nu) = -\frac{N}{M^{1/2}\omega_{\mathbf{q},\nu}^{3/2}} \int d\Omega \left(\mathbf{e}_{\mathbf{q},\nu} \cdot \nabla v(\mathbf{r})\right) \psi^{*}_{\mathbf{k},n,s}(\mathbf{r})\psi_{\mathbf{k}-\mathbf{q},n',s}(\mathbf{r})$$

(2.10)

If the matrix element γ depends only on the momentum transfer $\mathbf{q}$ so that

$$\gamma_{n,n'}(\mathbf{q},\mathbf{k},\nu) = \gamma(\mathbf{q})$$

we call the interaction the Fröhlich one.

2.1.2 The BCS approximation for the electron-electron interaction

The Fröhlich hamiltonian, Eq.(2.9) describes the emission and absorption of phonons by electrons tunneling in a band. In second order it generates an interaction between two electrons via the phonon exchange. This second order term is responsible for an instability of the Fermi sea versus pairing and should be treated not perturbatively. One can eliminate from the hamiltonian the linear term in the electron-phonon coupling γ retaining the essential second order one with a canonical transformation S. Introducing a new set of many-particle functions:

$$|\tilde{N}>= exp(S)|N>$$

(2.11)

one can rewrite the Schrödinger equation :

$$\tilde{H}|\tilde{N}>= E|\tilde{N}>$$

(2.12)

with

$$\tilde{H} = exp(S)Hexp(-S), \qquad (2.13)$$

and choose S in a form, which allows us to eliminate off-diagonal terms in phonon variables. Separating the first order contribution S_1

$$S = S_1 + S_2 \qquad (2.14)$$

with S_2 containing quadratic and higher powers of γ, one obtains with accuracy up to γ^2:

$$\tilde{H} = H + [S_1 H_0] + [S_1 H_{e-ph}] + [S_2 H_0] + \frac{1}{2}[S_1[S_1 H_0]]. \qquad (2.15)$$

To eliminate the first order term we determine S_1 from the condition:

$$H_{e-ph} + [S_1 H_0] = 0 \qquad (2.16)$$

which gives:

$$S_1 = \frac{1}{\sqrt{2N}} \sum_{k,q,\nu,s} \frac{\gamma(q,k,\nu)\omega_{q,\nu}c^{\dagger}_{k+q,s}c_{k,,s}d_{q,\nu}}{\xi_k - \xi_{k+q} + \omega_{q,\nu}} - h.c \qquad (2.17)$$

and

$$\tilde{H} = H_0 + \frac{1}{2}[S_1 H_{e-ph}] + [S_2 H_0]. \qquad (2.18)$$

Here and below we suppress the band index n assuming for simplicity that the matrix $\gamma_{n,n'}$ is diagonal.

S_2 can be determined so as to compensate off-diagonal matrix elements of the operator $[S_1 H_{e-ph}]$ retaining the terms containing only operators $d^{\dagger}_{q,\nu}d_{q,\nu}$ diagonal in phonon variables. As a result :

$$\tilde{H} = H_0 + \frac{1}{2}[S_1 H_{e-ph}]_{diag} \qquad (2.19)$$

For zero or sufficiently low temperature the number of phonons is small, $\langle d^{\dagger}_{q,\nu}d_{q,\nu}\rangle \simeq 0$ and only zero-point quantum fluctuations give rise to the interaction in this case:

$$\frac{1}{2}[S_1 H_{e-ph}] = \frac{1}{2N} \sum_{k,k',q,s,s'\nu} \frac{\gamma^2(q,k,\nu)\omega^3_{q,\nu}c^{\dagger}_{k+q,s}c^{\dagger}_{k'-q,,s'}c_{k',s'}c_{k,s}}{(\xi_k - \xi_{k+q})^2 - \omega^2_{q,\nu}}$$

$$(2.20)$$

One can see from Eq.(2.20) that two electrons attract each other if the energy transfer during the collision is small enough:

$$|\xi_{\mathbf{k}} - \xi_{\mathbf{k+q}}| < \omega_{\mathbf{q},\nu} \qquad (2.21)$$

Because the momentum transfer in the electron-phonon collisions is of order of the electron momentum itself, which in metals is comparable with the inverse lattice constant $1/a$, the characteristic phonon frequency is of the order of the Debye frequency ω_D. Therefore following BCS instead of Eq.(2.20) one can introduce a simple approximation for the interaction, which describes the essential physics of a superconductor:

$$V_{BCS} = -2E_p \qquad (2.22)$$

if the condition Eq.(2.21) is met with $\omega_{\mathbf{q},\nu} = \omega_D$, and $V_{BCS} = 0$ otherwise. Here

$$E_p = \frac{1}{2N} \sum_{\mathbf{q},\nu} \gamma^2(\mathbf{q},\nu)\omega_{\mathbf{q},\nu} \qquad (2.23)$$

is a so-called polaronic level shift, which is a characteristic potential energy of an electron in a field created by a lattice deformation. For simplicity we assume that the matrix element γ depends only on the momentum transfer $\mathbf{q}$.

Thus the hamiltonian of the system can be written as:

$$\tilde{H} = \sum_{\mathbf{k},s} \xi_{\mathbf{k}} c^\dagger_{\mathbf{k},s} c_{\mathbf{k},s} + \frac{1}{2} \sum_{\mathbf{k},\mathbf{K},s,s'} c^\dagger_{\mathbf{k},s} c^\dagger_{-\mathbf{k+K},s'} \hat{\Delta}_{ss'}(\mathbf{k},\mathbf{K}) \qquad (2.24)$$

with

$$\hat{\Delta}_{ss'}(\mathbf{k},\mathbf{K}) = \sum_{\mathbf{k}'} V_{BCS} c_{-\mathbf{k}'+\mathbf{K},s'} c_{\mathbf{k}',s} \qquad (2.25)$$

2.1.3   Ground state and excitations

To diagonalize the hamiltonian Eq.(2.24) BCS applied the same approximation as Bogoliubov used for a weakly interacting Bose-gas (Chapter 1). Pairs of electrons are bosons, which condense in a ground state with zero momentum of a pair, so that $K = 0$. The number of condensed pairs is macroscopically large, which means that a pair annihilation and creation operators $\hat{\Delta}$, $\hat{\Delta}^\dagger$ with $K = 0$

can be replaced by numbers, which for an open system are their averages:

$$\hat{\Delta}_{ss'}(\mathbf{k}, \mathbf{K}) \simeq (1 - \delta_{s,s'})\delta_{\mathbf{K},0}\Delta_{\mathbf{k}} \tag{2.26}$$

where

$$\Delta_{\mathbf{k}} = \sum_{\mathbf{k}'} V_{BCS}\langle c_{-\mathbf{k}',\downarrow}c_{\mathbf{k}',\uparrow}\rangle \tag{2.27}$$

As a rule only electrons with opposite spins are paired, and $\hat{\Delta}_{ss'}$ is off-diagonal. With the substitution Eq.(2.26) the reduced hamiltonian turns out to be quadratic in electron operators and can be diagonalised:

$$
\begin{aligned}
\tilde{H} &= \sum_{\mathbf{k}} \left[\xi_{\mathbf{k}} \left(c_{\mathbf{k},\uparrow}^{\dagger}c_{\mathbf{k},\uparrow} + c_{-\mathbf{k},\downarrow}^{\dagger}c_{-\mathbf{k},\downarrow} \right) + \Delta_{\mathbf{k}}c_{\mathbf{k},\uparrow}^{\dagger}c_{-\mathbf{k},\downarrow}^{\dagger} + \Delta_{\mathbf{k}}^{*}c_{-\mathbf{k},\downarrow}c_{\mathbf{k},\uparrow} \right] \\
&\quad - \frac{|\Delta|^2}{V_{BCS}}.
\end{aligned}
\tag{2.28}
$$

The last term in Eq.(2.28) is added to provide the equality of the ground state energies of the 'exact' hamiltonian, Eq.(2.24), and reduced one, Eq.(2.28). The linear transformation:

$$c_{\mathbf{k},\uparrow} = u_{\mathbf{k}}\alpha_{\mathbf{k}} + v_{\mathbf{k}}\beta_{\mathbf{k}}^{\dagger} \tag{2.29}$$

$$c_{-\mathbf{k},\downarrow} = u_{\mathbf{k}}\beta_{\mathbf{k}} - v_{\mathbf{k}}\alpha_{\mathbf{k}}^{\dagger} \tag{2.30}$$

reduces the problem of correlated electrons to an ideal Fermi-gas, consisting of two types of noninteracting quasiparticles α and β, which are fermions if the coefficients of the Bogoliubov transformation are chosen in the form:

$$u_{\mathbf{k}}^2 = \frac{1}{2}\left(1 + \frac{\xi_{\mathbf{k}}}{\epsilon_{\mathbf{k}}}\right) \tag{2.31}$$

$$v_{\mathbf{k}}^2 = \frac{1}{2}\left(1 - \frac{\xi_{\mathbf{k}}}{\epsilon_{\mathbf{k}}}\right) \tag{2.32}$$

and

$$u_{\mathbf{k}}v_{\mathbf{k}} = -\frac{\Delta_{\mathbf{k}}}{2\epsilon_{\mathbf{k}}} \tag{2.33}$$

The transformed hamiltonian is of the form:

$$\tilde{H} = E_0 + \sum_{\mathbf{k}} \epsilon_{\mathbf{k}}\left(\alpha_{\mathbf{k}}^{\dagger}\alpha_{\mathbf{k}} + \beta_{\mathbf{k}}^{\dagger}\beta_{\mathbf{k}}\right) \tag{2.34}$$

with the ground state energy

$$E_0 = 2 \sum_{\mathbf{k}} \left(\xi_{\mathbf{k}} v_{\mathbf{k}}^2 + \Delta_{\mathbf{k}} u_{\mathbf{k}} v_{\mathbf{k}} \right) - \frac{\Delta^2}{V_{BCS}} \tag{2.35}$$

and the energy of a one-particle excitation

$$\epsilon_{\mathbf{k}} = \sqrt{\xi_{\mathbf{k}}^2 + \Delta_{\mathbf{k}}^2} \tag{2.36}$$

A selfconsistent order parameter Δ which is proportional to the square root of the condensate density is determined from the fundamental BCS equation (2.27). After the substitution Eq.(2.29,30)

$$\Delta_{\mathbf{k}} = -\sum_{\mathbf{k}'} V_{BCS} \frac{\Delta_{\mathbf{k}'}}{\epsilon_{\mathbf{k}'}} \left(1 - 2 f_{\mathbf{k}'} \right) \tag{2.37}$$

where $f_{\mathbf{k}} = \langle \alpha_{\mathbf{k}}^{\dagger} \alpha_{\mathbf{k}} \rangle = \langle \beta_{\mathbf{k}}^{\dagger} \beta_{\mathbf{k}} \rangle$ is the quasiparticle distribution function. Unlike the case of bare electrons the average number of quasiparticles is not fixed. Therefore in thermal equilibrium their chemical potential is zero and:

$$f_{\mathbf{k}} = \frac{1}{exp(\epsilon_{\mathbf{k}}/T) + 1}. \tag{2.38}$$

One can replace the summation by an integral, using the definition of the density of states (DOS) $N(\xi)$ in a Bloch band:

$$N(\xi) = \sum_{\mathbf{k}} \delta(\xi - \xi_{\mathbf{k}}) \tag{2.39}$$

The limits are determined by the region $-\omega_D$ to $+\omega_D$ within which $V_{BCS} = -2E_p$. If $\omega_D << \mu$ DOS is practically constant in the narrow energy region around the Fermi-energy: $N(\xi) \simeq N(0)$. Finally the fundamental equation becomes:

$$\frac{\Delta}{\lambda} = \int_0^{\omega_D} \frac{d\xi \Delta \tanh \frac{\sqrt{\xi^2 + \Delta^2}}{T}}{\sqrt{\xi^2 + \Delta^2}} \tag{2.40}$$

with the electron-phonon coupling constant $\lambda = 2E_p N(0)$.

The order parameter Δ turns out to be independent of the momentum in a region $|\xi_{\mathbf{k}}| < \omega_D$ and equal zero outside this region. Below some critical temperature T_c there are two solution of

Eq.(2.40) one of them is trivial $\Delta = 0$. Above T_c the only solution is a normal state one, $\Delta = 0$. The system prefers to be in a super-conducting condensed state below T_c if the condensation energy E_c, which is the energy referred to the normal state, is negative. At $T = 0$ the ground state is a quasiparticle vacuum, $f_{\mathbf{k}} = 0$ and:

$$E_c = E_0 - 2 \sum_{\xi_{\mathbf{k}} < 0} \xi_{\mathbf{k}} \qquad (2.41)$$

Using the definition of E_0 , Eq.(2.35) , one obtains:

$$E_c = 2N(0) \int_0^{\omega_D} d\xi \left(\xi - \frac{\xi^2 + \Delta^2(0)/2}{\sqrt{\xi^2 + \Delta^2(0)}} \right) \qquad (2.42)$$

with $\Delta(0)$ the order parameter at $T = 0$. The integral in Eq.(2.42) gives a negative value

$$E_c \simeq -\frac{\Delta^2(0)N(0)}{2}. \qquad (2.43)$$

The assumption $\Delta << \omega_D$ has been used.

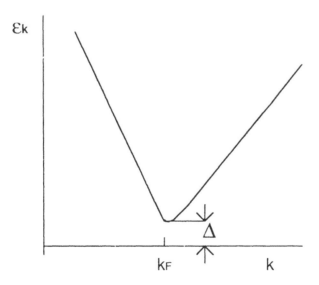

Fig.2.1. Excitation spectrum of the BCS superconductor.

Far away from the Fermi-surface the quasiparticles α and β are electrons with spin 'up' and 'down', correspondingly, if $\xi_{\mathbf{k}} > 0$ and

holes if $\xi_k < 0$. In the vicinity of the Fermi-surface they are mixture of both and their energy dispersion ϵ_k is remarkably different
from that of noninteracting electrons and holes (see Fig. 2.1). The
quasiparticle energy spectrum satisfies the Landau criterion for
superfluidity with the critical velocity $v_c \simeq v_F \Delta / \mu$, Fig.2.1. The
distribution of bare electrons is of the form ($T = 0$)

$$n_k = \langle c_{k,\uparrow}^\dagger c_{k,\uparrow} \rangle = v_k^2 \tag{2.44}$$

which has a zero step $Z = 0$ different from the Fermi distribution
at $T = 0$ with $Z = 1$, Fig.2.2. This is a clear manifestation of the
breakdown of the Fermi-liquid description of attractive fermions
at low temperature.

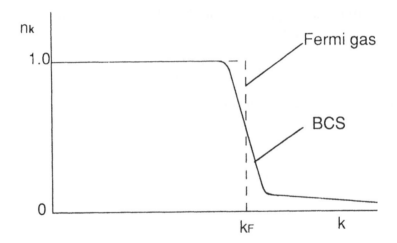

Fig.2.2. Distribution of electrons at $T = 0$.

2.2 Hallmarks of the BCS superconductivity

There are several hallmark indicators of the BCS superconductivity which can be checked experimentally.

2.2.1 The Meissner-Ochsenfeld effect

To demonstrate this effect in the BCS superconductor one can
apply the perturbation theory for the linear interaction of electrons

with a vector potential $\mathbf{A}(\mathbf{r})$

$$H_{int} = -\frac{e}{m} \sum_{\mathbf{k},\mathbf{q},s} (\mathbf{k} \cdot \mathbf{A_q}) c_{\mathbf{k}+\mathbf{q},s}^\dagger c_{\mathbf{k},s} \qquad (2.45)$$

where $\mathbf{A_q}$ is the Fourier component of $\mathbf{A}(\mathbf{r})$. This form of the interaction with a magnetic field follows from the velocity operator $(-i\nabla - e\mathbf{A}(\mathbf{r}))/m$ in the effective mass (m) approximation for the band energy dispersion and in the Coulomb gauge with $div\,\mathbf{A}(\mathbf{r}) = 0$. The field distribution in a sample is determined by the average of the current density operator, which follows as the symmetrised form of the velocity operator in the second -quantization

$$\hat{\mathbf{j}}(\mathbf{r}) = \hat{\mathbf{j}}_p(\mathbf{r}) + \hat{\mathbf{j}}_d(\mathbf{r}) \qquad (2.46)$$

where the paramagnetic contribution

$$\hat{\mathbf{j}}_p(\mathbf{r}) = \frac{e}{2m} \sum_{\mathbf{k},\mathbf{q},s} c_{\mathbf{k}+\mathbf{q},s}^\dagger c_{\mathbf{k},s} (2\mathbf{k} + \mathbf{q})\, exp(-i\mathbf{q} \cdot \mathbf{r}) \qquad (2.47)$$

and

$$\hat{\mathbf{j}}_d(\mathbf{r}) = -\frac{e^2}{m} \sum_{\mathbf{k},\mathbf{q},s} c_{\mathbf{k}+\mathbf{q},s}^\dagger c_{\mathbf{k},s} \mathbf{A}(\mathbf{r})\, exp(-i\mathbf{q} \cdot \mathbf{r}) \qquad (2.48)$$

the diamagnetic part. The perturbed many-particle state $|\tilde{N}\rangle_1$ within a linear approximation in $\mathbf{A}$ is given by

$$|\tilde{N}\rangle_1 = |n_\alpha, n_\beta\rangle + \sum_{n'_\alpha, n'_\beta} |n'_\alpha, n'_\beta\rangle \frac{\langle n'_\alpha, n'_\beta | H_{int} | n_\alpha, n_\beta \rangle}{E_{n_\alpha,n_\beta} - E_{n'_\alpha,n'_\beta}} \qquad (2.49)$$

where $|n_\alpha, n_\beta\rangle$ and E_{n_α,n_β} are eigenstates and eigenvalues of the reduced hamiltonian $\tilde{H}$, correspondingly.

This expression together with the Bogoliubov transformation, Eq.(2.29,30), yields for the average current density at $T = 0$

$$\begin{aligned} \mathbf{j}_p(\mathbf{r}) &= \frac{e^2}{m^2} \sum_{\mathbf{k},\mathbf{q}} \frac{2u_{\mathbf{k}+\mathbf{q}}v_{\mathbf{k}}(2\mathbf{k} + \mathbf{q})(\mathbf{k} \cdot \mathbf{A_q})\exp(-i\mathbf{q} \cdot \mathbf{r})}{\epsilon_{\mathbf{k}+\mathbf{q}} + \epsilon_{\mathbf{k}}} \\ &\times \left(u_{\mathbf{k}+\mathbf{q}}v_{\mathbf{k}} - u_{\mathbf{k}}v_{\mathbf{k}+\mathbf{q}} \right) \end{aligned} \qquad (2.50)$$

$$\mathbf{j}_d = -\frac{ne^2}{m} \mathbf{A}(\mathbf{r}) \qquad (2.51)$$

Let us assume that the magnetic field varies with a characteristic length λ_H, which is large compared with the coherence length ξ. In this case one can take a limit $q \to 0$ in Eq.(2.50). In this limit $u_{k+q}v_k - u_k v_{k+q} \simeq 0$ while the denominator remains finite $\epsilon_{k+q} + \epsilon_k > 2\Delta(0)$. Therefore the paramagnetic contribution vanishes and we are left with the London equation

$$\mathbf{j(r)} = -\frac{ne^2}{m}\mathbf{A(r)} \qquad (2.52)$$

which together with the Maxwell equation yields $\lambda_H = \sqrt{m/4\pi ne^2}$. In the opposite limit $\lambda_H < \xi$ the Pippard (1953) nonlocal modification of the London equation takes place.

On the contrary in the normal state the denominator in Eq.(2.50) turns out to be zero at the Fermi level, and paramagnetic current compensates the diamagnetic one, assuring a zero total current in a constant magnetic field.

2.2.2 BCS gap, critical temperature and tunneling

It follows from the above consideration that the BCS superconductivity is linked with the order parameter Δ, which is defined as a gap in the quasiparticle spectrum, Fig.2.1 for a homogeneous system. The value of the gap at $T = 0$ should be of the order of T_c. In fact the BCS theory predicts a universal ratio $2\Delta(0)/T_c \simeq 3.5$ as follows from the fundamental Eq (2.40). At $T = 0$ the nontrivial solution is determined from

$$\frac{1}{\lambda} = \int_0^{\omega_D} \frac{d\xi}{\sqrt{\xi^2 + \Delta^2(0)}}, \qquad (2.53)$$

Integration yields

$$\Delta(0) \simeq 2\omega_D exp(-\frac{1}{\lambda}) \qquad (2.54)$$

for $\lambda << 1$, the limit to which the theory is applied. This is a remarkable result, which demonstrates the instability of the Fermi-liquid for *any value* of the attraction λ. The exponent in Eq.(2.54) cannot be expanded in a series of λ. Therefore the superconducting ground state cannot be derived by using a perturbation theory.

At $T = T_c$ the gap is zero, and T_c is determined by

$$\frac{1}{\lambda} = \int_0^{\omega_D} \frac{d\xi \tanh\left(\frac{\xi}{2T_c}\right)}{\xi}, \tag{2.55}$$

Integrating by parts and replacing the upper limit by infinity we have

$$T_c = \frac{2e^C \omega_D}{\pi} exp(-\frac{1}{\lambda}) \tag{2.56}$$

where $C \simeq 0.577$ is the Euler constant. The coefficient in Eq.(2.56) is $\simeq 1.14$ and

$$\frac{2\Delta(0)}{T_c} \simeq 3.5 \tag{2.57}$$

At low temperature, $T << T_c$ one can expand

$$\tanh \frac{\sqrt{\xi^2 + \Delta^2(T)}}{2T} \simeq 1 - exp\left(-\frac{\Delta(0)}{T} - \frac{\xi^2}{2T\Delta(0)}\right) \tag{2.58}$$

and

$$\Delta(T) \simeq \Delta(0) + \Delta_1(T) \tag{2.59}$$

where $\Delta_1(T) << \Delta(0)$. The substitution of these expansions into Eq.(2.40) yields an exponentially small correction $\Delta_1(T)$ given by

$$\Delta_1(T) = -\sqrt{2\pi T\Delta(0)}exp(-\frac{\Delta(0)}{T}) \tag{2.60}$$

In the vicinity of T_c where $(T_c - T)/T_c << 1$ the gap is small compared with temperature. However direct expansion in powers of Δ cannot be applied to Eq.(2.40). Instead it is convenient to use

$$\frac{\tanh x}{x} = \sum_{n=-\infty}^{\infty} \frac{1}{x^2 + [\pi(n + 1/2)]^2} \tag{2.61}$$

and

$$\frac{1}{\lambda} = 2T \sum_n \int_0^{\Theta} \frac{d\xi}{\xi^2 + \omega_n^2 + \Delta^2} \tag{2.62}$$

where $\omega_n = \pi T(2n + 1)$ are the so-called Matsubara frequencies, $n = 0, \pm1, \pm2...$. The last equation can be expanded in powers of Δ

$$ln\frac{T_c}{T} = 2\Delta^2(T)T_c \sum_n \int_0^{\infty} \frac{d\xi}{(\xi^2 + \omega_n^2)^2} \tag{2.63}$$

and

$$\Delta(T) = \pi T_c[8/7\zeta(3)]^{1/2}\sqrt{1 - \frac{T}{T_c}} \simeq 3.06T_c\sqrt{1 - \frac{T}{T_c}} \qquad (2.64)$$

There is a discontinuity in the temperature derivative of Δ at T_c which leads to a jump of the heat capacity (see 2.2.4). Therefore the BCS superconducting transition is of the second kind.

The most direct way to measure the gap is a tunneling experiment in which one applies the voltage V to a thin dielectric layer between a normal metal and a superconductor, Fig.2.3 (Giaever(1960)).

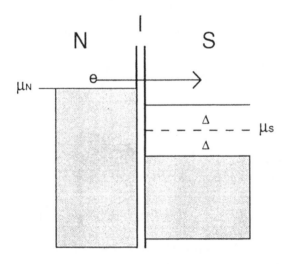

Fig.2.3. Tunneling from a normal metal (N) to a superconductor (S) through a dielectric barrier. Dashed areas correspond to occupied states.

The current through the dielectric is proportional to the number of electrons tunneling under the barrier per second, which can be calculated with the Fermi golden rule

$$I(V) \sim \sum_{k(\xi_k<0),k'} T^2_{k,k'}\delta(\epsilon_{k'} - \xi_k - eV) \qquad (2.65)$$

with the matrix element $T_{k,k'}$ practically independent of the momentum k in a normal metal and k' in a superconductor (if the voltage is not very high $eV \sim \Delta << \mu$). The δ-function in Eq.(2.65) takes into account the difference eV in the 'left' and 'right' chemical potentials. Replacing the summation by the integration with the constant 'left' and 'right' DOS we obtain

$$I(V) \sim \int_{-\infty}^{eV} d\xi \int_{-\infty}^{+\infty} d\xi' \delta(\sqrt{(\xi')^2 + \Delta^2(T)} - \xi) \qquad (2.66)$$

The gap appears more pronounced in the conductance $\sigma = dI/dV$

$$\sigma \sim \int_{-\infty}^{+\infty} d\xi' \delta(\sqrt{(\xi')^2 + \Delta^2(T)} - eV) \qquad (2.67)$$

which by definition is a density of quasiparticle states of a superconductor. From Eq.(2.67) we have

$$\frac{\sigma}{\sigma_N} = \frac{eV}{\sqrt{(eV)^2 - \Delta^2(T)}} \qquad (2.68)$$

with σ_N the conductance of the barrier above T_c. There is no current if $eV < \Delta$ because there are no states inside the gap in a superconductor, Fig.2.3. Just above the threshold $eV = \Delta$ the conductance has a maximum because the quasiparticle density of states diverges at $\epsilon = \Delta$. The typical experimental ratio σ/σ_N as a function of the voltage and the temperature dependence of the gap, determined by tunneling, follow rather well the BCS prediction in classical low-temperature superconductors.

2.2.3 Isotope effect

The nature of the electron-electron attraction in solids can be checked by isotope substitution, when the ion mass M is varied without any change of the electronic configuration of an ion. There are two parameters in the BCS expression for T_c, Eq.(2.56), which depend on the mechanism of the interaction. The characteristic phonon frequency ω_D is proportional to $1/\sqrt{M}$, as a frequency of any harmonic oscillator. On the other hand the coupling constant λ is independent on the ion mass. This follows from the expression Eq.(2.10) for the matrix element

$$\gamma \sim (M\omega^3)^{-1/2} \sim M^{1/4} \qquad (2.69)$$

The product $\gamma^2\omega$ does not depend on M and the same is true for λ. That is why

$$\alpha = -\frac{dlnT_c}{dlnM} = 0.5 \tag{2.70}$$

if the electron-phonon interaction is responsible for superconductivity. In fact the isotope exponent α may be lower than 0.5 in the BCS superconductor because of the Coulomb repulsion and the anharmonicity of phonons. Therefore any finite value of α measured experimentally shows that phonons are involved in the pairing mechanism. However the absence or a small isotope effect does not mean that the electron-phonon interaction is irrelevant for superconductivity.

2.2.4 Heat capacity

Because of the Pauli principle only electrons near the Fermi-surface can absorb heat in a metal. The number of these electrons is proportional to temperature. Therefore their entropy and heat capacity C_e in the normal state are linear in temperature. Because of the gap the temperature dependence of C_e changes drastically in the superconducting state. The entropy of an ideal Fermi-gas of qusiparticles is

$$S = -2\sum_{\mathbf{k}}[f_{\mathbf{k}}lnf_{\mathbf{k}} + (1 - f_{\mathbf{k}})ln(1 - f_{\mathbf{k}})] \tag{2.71}$$

where the distribution function is defined with Eq.(2.38). Integrating by parts we obtain

$$S = \frac{4N(0)}{T}\int_0^\infty d\xi \frac{2\xi^2 + \Delta^2(T)}{\sqrt{\xi^2 + \Delta^2(T)}(1 + exp^{\frac{\sqrt{\xi^2 + \Delta^2(T)}}{T}})} \tag{2.72}$$

At low temperature $S \sim e^{-\Delta(0)/T}$ because the number of quasiparticles is exponentially small, and the same is valid for C_e. Near T_c one can expand the integral Eq.(2.72) in powers of Δ

$$S \simeq S_N - 2N(0)\frac{\Delta^2(T)}{T_c}, \tag{2.73}$$

with the normal state entropy $S_N = 4\pi^2 N(0)T/3$. As a result the heat capacity $C_e = TdS/dT$ has a jump at T_c. Just above T_c

$$C_e = C_N = \frac{4\pi^2}{3}N(0)T_c \qquad (2.74)$$

if $T = T_c + 0$ and just below

$$C_e = C_N + \frac{16\pi^2}{7\zeta(3)}N(0)T_c \qquad (2.75)$$

if $T = T_c - 0$. Here $\zeta(3) \simeq 1.202$. The relative value of the jump

$$\frac{C_e(T_c - 0) - C_e(T_c + 0)}{C_N} = \frac{12}{7\zeta(3)} \simeq 1.43 \qquad (2.76)$$

which is in a perfect agreement with the experimental results for classical superconductors.

2.2.5 Sound attenuation

The interaction of the ultrasound wave with electrons is described by

$$H_{int} = V(q)e^{i\nu t}\sum_{\mathbf{k},s} c^\dagger_{\mathbf{k},s}c_{\mathbf{k-q},s} + h.c., \qquad (2.77)$$

where $q = \nu/s$ is the wave-vector, ν the frequency and s the sound velocity. $V(q)$ is proportional to the sound amplitude.

Applying the Bogoliubov transformation there are four terms in the interaction, Eq.(2.77). Two of them correspond to annihilation and creation of two different quasiparticles and the other two correspond to their scattering. For low enough frequency $\nu << \Delta_0 \simeq T_c$ only the scattering of quasiparticles is relevant:

$$H_{int} \sim \sum_{\mathbf{k}} M^{(-)}_{\mathbf{k},\mathbf{q}}\left(\alpha^\dagger_\mathbf{k}\alpha_{\mathbf{k-q}} + \beta^\dagger_\mathbf{k}\beta_{\mathbf{k-q}}\right), \qquad (2.78)$$

where

$$M^{(-)}_{\mathbf{k},\mathbf{q}} = u_\mathbf{k}u_{\mathbf{k-q}} - v_\mathbf{k}v_{\mathbf{k-q}} \qquad (2.79)$$

is a so-called coherence factor.

The rate of the absorption of sound is given by

$$W_{abs} \sim \sum_{\mathbf{k}}(M^{(-)}_{\mathbf{k},\mathbf{q}})^2 f_{\mathbf{k-q}}(1 - f_\mathbf{k})\delta(\epsilon_{\mathbf{k-q}} - \epsilon_\mathbf{k} + \nu) \qquad (2.80)$$

and the emission rate

$$W_{emi} \sim \sum_{\mathbf{k}} (M_{\mathbf{k,q}}^{(-)})^2 f_{\mathbf{k-q}} (1 - f_{\mathbf{k}}) \delta(\epsilon_{\mathbf{k-q}} - \epsilon_{\mathbf{k}} - \nu) \qquad (2.81)$$

The sound attenuation Γ, which is the difference of these two, is given by

$$\Gamma \sim \sum_{\mathbf{k}} (M_{\mathbf{k,q}}^{(-)})^2 (f_{\mathbf{k}} - f_{\mathbf{k-q}}) \delta(\epsilon_{\mathbf{k-q}} - \epsilon_{\mathbf{k}} - \nu) \qquad (2.82)$$

For $\nu << T, \Delta$ we have $f_{\mathbf{k}} - f_{\mathbf{k-q}} \simeq \nu \partial f / \partial \epsilon$. Replacing the integration over k and over the angle between $\mathbf{k}$ and $\mathbf{q}$ for the integration over $\xi = \xi_{\mathbf{k}}$ and $\xi' = \xi_{\mathbf{k-q}}$ correspondingly, we obtain

$$\Gamma \sim - \int d\xi \int d\xi' \, (u(\xi)u(\xi') - v(\xi)v(\xi'))^2 \frac{\partial f}{\partial \epsilon} \delta(\epsilon - \epsilon'), \qquad (2.83)$$

with $\epsilon = \sqrt{\xi^2 + \Delta^2}$ and $\epsilon' = \sqrt{\xi'^2 + \Delta^2}$.

Only transitions with $\xi = \xi'$ contribute:

$$\Gamma \sim - \int_{\Delta}^{\infty} d\epsilon \left(\frac{\partial \xi}{\partial \epsilon} \right)^2 \left(u^2(\xi) - v^2(\xi) \right)^2 \frac{\partial f}{\partial \epsilon} \qquad (2.84)$$

One can see that the large density of quasiparticle states $\frac{\partial \xi}{\partial \epsilon} = \epsilon / \sqrt{\epsilon^2 - \Delta^2}$ is canceled by the small coherence factor in the integral, Eq.(2.84):

$$\Gamma \sim - \int_{\Delta}^{\infty} d\epsilon \frac{\epsilon^2 - \Delta^2}{\epsilon^2 - \Delta^2} \frac{\partial f}{\partial \epsilon} = \frac{1}{exp(\frac{\Delta}{T}) + 1} \qquad (2.85)$$

The sound attenuation in the superconducting state depends on temperature exponentially, Fig.2.4. Using its ratio to the normal state attenuation one can measure the temperature dependence of the BCS gap:

$$\frac{\Gamma_s}{\Gamma_n} = \frac{2}{exp(\frac{\Delta(T)}{T}) + 1} \qquad (2.86)$$

2.2.6 Nuclear spin relaxation rate

The other powerful method of determining $\Delta(T)$ is the measurement of the line-width of the nuclear magnetic resonance (NMR), which depends on the inverse time of the relaxation of nuclear magnetic moment due to the spin-flip scattering of carriers on nuclear, $1/T_1$. This scattering is described by the hyperfine interaction of the nuclear with the electron spin:

$$H_{int} \sim \sum_{k,k'} c^\dagger_{k',\downarrow} c_{k,\uparrow} + h.c.. \tag{2.87}$$

The NMR frequency is very small and the spin-flip scattering is practically elastic. That is why only the scattering of quasiparticles contribute to $1/T_1$:

$$H_{int} \sim \sum_k M^{(+)}_{k,k'} \left(\beta^\dagger_{k'} \alpha_k + \alpha^\dagger_k \beta_{k'} \right), \tag{2.88}$$

where

$$M^{(+)}_{k,k'} = u_k u_{k'} + v_k v_{k'} \tag{2.89}$$

is another coherence factor.

With the Fermi golden rule :

$$1/T_1 \sim \sum_{k,k'} (M^{(+)}_{k,k'})^2 f_k (1 - f_{k'}) \delta(\epsilon_k - \epsilon_{k'}) \tag{2.90}$$

Replacing summation on integration we obtain a divergent integral, which can be cut by some damping of excitations Γ (for example due to the inelastic electron-phonon scattering):

$$1/T_1 \sim -T \int_\Delta^\infty d\epsilon \frac{\epsilon^2 + \Delta^2}{\epsilon^2 - \Delta^2} \frac{\partial f}{\partial \epsilon} \tag{2.91}$$

Above T_c when $\Delta = 0$ the relaxation rate is proportional to T as for normal metals (Korringa law). Well below T_c

$$1/T_1 \sim e^{-\frac{\Delta}{T}} ln \frac{\Delta}{\Gamma}; \tag{2.92}$$

the relaxation rate is exponentially small like the sound attenuation. However just below T_c it has a maximum (Hebel-Slichter peak), Fig.2.4. The prediction of this rise of $1/T_1$ just below T_c is the most interesting and important feature of the BCS superconductor. This result is distinctly different from the result for the acoustic attenuation. They are different because the coherence factors are different:

$$M^{(+,-)} = (uu' \pm vv') \tag{2.93}$$

It is clear that a simple energy-gap form of a $two - fluid$ model could not account for the drop in the sound attenuation and simultaneously the rapid rise of $1/T_1$ near T_c. The observation by Hebel and Slichter (1959) gave one of the first evidence of the detailed nature of the pairing correlations in the BCS-superconductor.

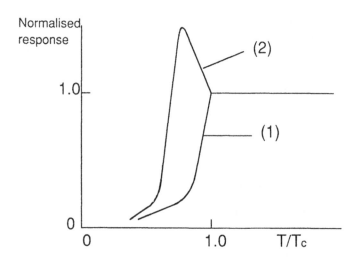

Fig.2.4. Temperature dependence of the sound attenuation and thermal conductivity (1) compared with the nuclear spin relaxation rate (2).

2.2.7   Thermal conduction of the BCS superconductor

Important information about the excitation spectrum of the superconducting state can also be obtained from the thermal conductivity (Geilikman (1958); Bardeen, Rickayzen and Tewordt (1959)). Quasiparticles contribute to the electron heat transfer in a superconductor. The electron heat flow is determined by the deviation $\tilde{f}$ of their distribution function from the equilibrium one f:

$$\mathbf{Q} = \sum_{\mathbf{k}} \tilde{f} \mathbf{v} \epsilon_{\mathbf{k}} \tag{2.94}$$

with the group velocity $\mathbf{v} = \frac{\partial \epsilon_{\mathbf{k}}}{\partial \mathbf{k}}$.

The distribution function obeys the Boltzmann equation, which for a small deviation from equilibrium has the form:

$$\mathbf{v}\frac{\partial f}{\partial \mathbf{r}} - \frac{\partial \epsilon_{\mathbf{k}}}{\partial \mathbf{r}}\frac{\partial f}{\partial \mathbf{k}} = -\frac{\tilde{f}}{\tau_{tr}^s} \quad (2.95)$$

where the transport relaxation rate for the elastic scattering is given by the Fermi-golden rule:

$$\frac{1}{\tau_{tr}^s} = \frac{1}{\tau_{tr}^n}\frac{|\xi|}{\epsilon} \quad (2.96)$$

In the superconducting state the transport relaxation rate is diminished compared with that in the normal state (n) because of the square of the coherence factor $M^{(-)}$ in the probability of the scattering. The second term on the left hand side of the Boltzmann equation takes into account a driving force which acts on a quasiparticle due to the temperature dependence of the excitation energy, $\epsilon = \sqrt{\xi^2 + \Delta^2(T(\mathbf{r}))}$.

The solution of Eq.(2.95)

$$\tilde{f} = \tau_{tr}^n \frac{\epsilon^2}{T|\xi|}\frac{\partial f}{\partial \epsilon}\mathbf{v}\cdot\nabla T \quad (2.97)$$

after substitution in Eq.(2.94) yields for the thermal conductivity $K = |\mathbf{Q}|/|\nabla T|$:

$$K_s \sim -\frac{1}{T}\int_\Delta^\infty d\epsilon\epsilon^2\frac{\partial f}{\partial \epsilon} \quad (2.98)$$

with the ratio to the normal one

$$\frac{K_s}{K_n} = \frac{\int_{\Delta/2T}^\infty dx x^2 sech^2(x)}{\int_0^\infty dx x^2 sech^2(x)} \quad (2.99)$$

The normalized electronic thermal conductivity drops exponentially with the temperature lowering below T_c, Fig. 2.4.

The BCS theory is a mean-field approximation, which is valid if the number of carriers in a volume occupied by a pair is large. As an example, in *aluminium* the size of a pair (coherence length at $T = 0$) is ten thousand times larger than the distance between electrons. That is why pairs in the BCS superconductor disappear above the critical temperature T_c of the superconducting phase transition and one-particle excitations are fermions. With the increasing critical temperature or the coupling constant λ the coherence length becomes smaller and one can expect deviations from the BCS behavior. At first stage deviations from the BCS theory arise in two ways: (1) the assumption of an effective instantaneous interaction between electrons does not provide an adequate representation of the retarded nature of the phonon induced interaction; (2) the damping rate becomes comparable with the quasiparticle energy. Both the retardation effect and the damping are taken into account by the Eliashberg extension of the BCS theory to the *intermediate* coupling limit, which we discuss in Chapter 3. However when λ becomes larger than unity the Fermi liquid is unstable because of the *polaron collapse* of the electron band. The bipolaron theory of strong coupling superconductors is discussed in Chapters 4 and 5.

Chapter 3

Intermediate coupling superconductivity

3.1 Electron-phonon interaction in a normal metal

If the electron-phonon interaction is strong the perturbation theory discussed above does not work. The interaction affects both the one-particle excitation spectrum of the normal phase and the BCS superconducting state. Following Migdal (1958) we first discuss the normal state.

Higher order terms of the perturbation expansion contribute when λ is large. The Green's function diagram technique is instrumental in the summation of the substantial part of them. The one-particle excitation spectrum is determined by poles of the Fourier component of the one-particle Green's function (GF):

$$G(\mathbf{k}, t) = -i \langle T_t c_{\mathbf{k}}(t) c_{\mathbf{k}}^\dagger \rangle, \tag{3.1}$$

with $c_{\mathbf{k}}(t) = exp(iHt) c_{\mathbf{k}} exp(-iHt)$ and $T_t A(t) B = \Theta(t) A(t) B - \Theta(-t) B A(t)$ for any fermionic operators A, B and with the opposite sign of the second term for any bosonic operators, $\Theta(x) = 1$ for $x > 0$ and zero otherwise. The Fourier component of GF for free fermions $(H = H_0)$ is given by

$$G^{(0)}(\mathbf{k}, \omega) = \frac{\Theta(\xi_{\mathbf{k}})}{\omega - \xi_{\mathbf{k}} + i0^+} + \frac{\Theta(-\xi_{\mathbf{k}})}{\omega - \xi_{\mathbf{k}} - i0^+}, \tag{3.2}$$

which is obtained by the direct integration of Eq.(3.1):

$$G^{(0)}(\mathbf{k},\omega) = \frac{1}{2\pi}\int_{-\infty}^{\infty} dt G^0(\mathbf{k},t)e^{i\omega t} \tag{3.3}$$

For interacting electrons the electron self-energy is introduced:

$$\Sigma(\mathbf{k},\omega) = (G^{(0)}(\mathbf{k},\omega))^{-1} - (G(\mathbf{k},\omega))^{-1} \tag{3.4}$$

In this way we find

$$G(\mathbf{k},\omega) = \frac{1}{\omega - \xi_{\mathbf{k}} - \Sigma(\mathbf{k},\omega)} \tag{3.5}$$

and a new quasiparticle spectrum renormalised by the interaction is given near the Fermi level by:

$$\tilde{E}_{\mathbf{k}} = \tilde{\mu} + v_F(k - k_F) + \delta E_{\mathbf{k}} \tag{3.6}$$

where

$$\delta E_{\mathbf{k}} = \Sigma(\mathbf{k}, \tilde{E}_{\mathbf{k}} - \tilde{\mu}) - \Sigma(k_F, 0) \tag{3.7}$$

and $\tilde{\mu} = \mu + \Sigma(k_F, 0)$ is the renormalised Fermi energy. The spectrum is assumed to be isotropic and the spin and band quantum numbers are deleted for simplicity, v_F and k_F are the Fermi velocity and momentum respectively.

One can also determine a Fourier component of the free phonon Green's function as:

$$D^{(0)}(\mathbf{q},\omega) = \frac{\omega_{\mathbf{q}}^2}{\omega^2 - \omega_{\mathbf{q}}^2 + i0^+} \tag{3.8}$$

and the phonon self-energy

$$\Pi(\mathbf{q},\omega) = (D^{(0)}(\mathbf{q},\omega))^{-1} - (D(\mathbf{q},\omega))^{-1} \tag{3.9}$$

A selfconsistent set of two integral equations for Σ and Π within the Migdal approximation has the form:

$$\Sigma(\mathbf{k},\epsilon) = \frac{i}{(2\pi)^4}\int d\mathbf{q}d\omega\gamma^2(\mathbf{q})\omega_{\mathbf{q}}G(\mathbf{k}-\mathbf{q},\epsilon-\omega)D(\mathbf{q},\omega) \tag{3.10}$$

$$\Pi(\mathbf{q},\omega) = -\frac{2i\gamma^2(\mathbf{q})\omega_{\mathbf{q}}}{(2\pi)^4}\int d\mathbf{k}d\epsilon G(\mathbf{k}+\mathbf{q},\epsilon+\omega)G(\mathbf{k},\epsilon) \tag{3.11}$$

The electron self-energy, determined with Eq.(3.10) is a result of the summation of a particular infinite set of diagrams, Fig.3.1a, which does not include any diagram with the intersection of two phonon (wavy) lines, Fig 3.1b, and the same for the phonon self-energy, Fig.3.1c.

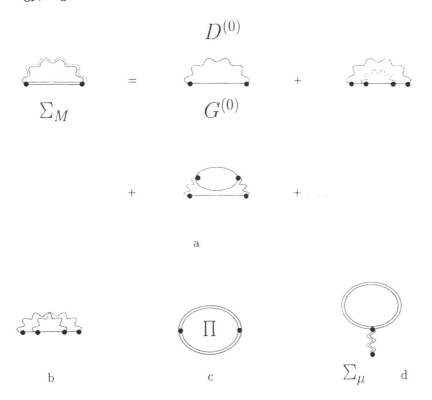

Fig.3.1. Electron (a,d) and phonon (c) self-energy in the Migdal approximation.

The additional 2 in the phonon self-energy is due to the contribution to the polarization loop, Fig 3.1c from two electron spins. The diagrams with intersections are small while the so-called adiabatic ratio is small:

$$\frac{\omega_{\mathbf{q}}}{\mu} << 1 \qquad (3.12)$$

In a simple metal with the Fermi energy of the order of $10000 K$

or larger and with the low Debye temperature $\omega_D \simeq 100K$ this is the case. The diagram for Σ with the closed electron loop, Fig.3.1d is frequency and momentum independent and according to the definition Eq. (3.6) should be included in the renormalised chemical potential. For acoustic phonons it is zero. The adiabatic condition Eq.(3.12) allows one to simplify the set of Eq.(3.10,11) further by replacing the exact GFs (solid lines) by free ones (thin lines) in the integrals. One can take for simplicity

$$\gamma^2(\mathbf{q})\omega_{\mathbf{q}} = 2E_p. \tag{3.13}$$

This is momentum independent, which is the case for long-wave acoustic phonons. As a result :

$$
\begin{aligned}
\delta E_{\mathbf{k}} &= \frac{2iE_p}{(2\pi)^4} \int d\mathbf{q}d\omega \left(G^{(0)}(\mathbf{k}-\mathbf{q}, \tilde{\xi}-w) - G^{(0)}(\mathbf{k}_F-\mathbf{q}, -w) \right) \\
&\times D^{(0)}(\mathbf{q}, \omega)
\end{aligned}
\tag{3.14}
$$

and

$$\Pi(\mathbf{q}, \omega) = -\frac{4iE_p}{(2\pi)^4} \int dkd\epsilon G^{(0)}(\mathbf{k}+\mathbf{q}, \epsilon+\omega)G^{(0)}(\mathbf{k}, \epsilon) \tag{3.15}$$

where $\tilde{\xi} = v_F(k - k_F) + \delta E_{\mathbf{k}}$.

The main contribution to the integral Eq.(3.14) comes from the momentum region close to the Fermi-surface:

$$|\mathbf{k} - \mathbf{q}| \simeq k_F \tag{3.16}$$

This makes it convenient to introduce a new variable $k' = |\mathbf{k} - \mathbf{q}|$ instead of the angle Θ between $\mathbf{k}$ and $\mathbf{q}$ and extend the integration to $\pm\infty$ with the variable $\xi = v_F(k' - k_F)$. Thus the angular integration in Eq.(3.14) becomes:

$$\int d\Theta sin\Theta (...) \sim \int_{\infty}^{\infty} d\xi \frac{\tilde{\xi}}{[\tilde{\xi} - w - \xi + i0^+ sgn(\xi)][w + \xi - i0^+ sgn(\xi)]} \tag{3.17}$$

This integral is non-zero only if $\tilde{\xi} > w > 0$ or $\tilde{\xi} < w < 0$. It is $-2\pi i$ in the first region and $2\pi i$ in the second one. Taking into account that $D^{(0)}$ is an even function of w one obtains:

$$\delta E_{\mathbf{k}} = \frac{2E_p}{(2\pi)^2 v_F} \int_0^{q_D} dqq \int_0^{|\tilde{\xi}|} dw sgn(\tilde{\xi}) \frac{w_q^2}{w^2 - w_q^2 + i0^+} \tag{3.18}$$

where q_D is the maximum (Debye) momentum of phonons of order of π/a and the phonon spectrum is assumed to be isotropic $\omega_{\mathbf{q}} = \omega_q$. The real and imaginary parts of Eq.(3.18) determine respectively the renormalised spectrum and the life time of quasi-particles:

$$Re\,(\delta E_{\mathbf{k}}) = \frac{E_p}{4\pi^2 v_F} \int_0^{q_D} dq q \omega_q \ln \left| \frac{\omega_q - \tilde{\xi}}{\omega_q + \tilde{\xi}} \right| \qquad (3.19)$$

$$Im\,(\delta E_{\mathbf{k}}) = \frac{E_p}{4\pi v_F} \int_0^{q_m} dq q \omega_q sgn(\tilde{\xi}) \qquad (3.20)$$

with $q_m = |\tilde{\xi}|/s$ if $|\tilde{\xi}| < \omega_D$ and $q_m = q_D$ if $|\tilde{\xi}| > \omega_D$. Here $s = \omega_D/q_D$ is the sound velocity, $\omega_q = sq$. For the excitations far away from the Fermi surface with $|\tilde{\xi}| >> \omega_D$

$$Re\,(\delta E_{\mathbf{k}}) = -\lambda \frac{\omega_D}{2\tilde{\xi}} \qquad (3.21)$$

where $\lambda = E_p q_D^2 / 4\pi^2 v_F$ is the same as the BCS coupling constant, determined in Eq.(2.40) because for a half-filled band with $k_F = q_D/2$ the density of states on the Fermi-level $N(0) = q_D^2/8\pi^2 v_F$. For low-energy excitations with $|\tilde{\xi}| << \omega_D$

$$Re\,(\delta E_{\mathbf{k}}) = -\lambda \tilde{\xi} \qquad (3.22)$$

which means an increase of the effective mass of the excitation due to the electron-phonon interaction

$$\tilde{\xi} = \frac{k_F}{m^*}(k - k_F) \qquad (3.23)$$

The renormalised effective mass

$$m^* = (1 + \lambda)m. \qquad (3.24)$$

Thus the excitation spectrum of a metal has two different regions with two different values of the effective mass. The thermodynamic properties of a metal at low temperature $T << \omega_D$ involve m^*, but the optical properties in a frequency range $\nu >> \omega_D$ are determined by the high-energy excitations, where according to Eq.(3.21) corrections are small and mass is equal to the band mass m.

The damping has just the opposite behavior. The integral Eq.(3.20) yields

$$Im\,(\delta E_{\mathbf{k}}) = \frac{sgn(\tilde{\xi})\pi\lambda\omega_D}{3} \tag{3.25}$$

if $|\tilde{\xi}| > \omega_D$, and

$$Im\,(\delta E_{\mathbf{k}}) = \frac{sgn(\tilde{\xi})\pi\lambda|\tilde{\xi}|^3}{3\omega_D^2} \tag{3.26}$$

for $|\tilde{\xi}| << \omega_D$.

These expressions describe the rate of decay of the quasiparticles due to emission of phonons. In the immediate neighborhood of the Fermi surface $|\tilde{\xi}| << \omega_D$ the decay is small compared with the quasiparticle energy $|\tilde{\xi}|$ even for a relatively strong coupling $\lambda \sim 1$, so that the concept of well-defined quasiparticles has a definite meaning. Hence within the Migdal approximation the electron-phonon interaction does not destroy the Fermi-liquid behavior of electrons. The Pauli exclusion principle is responsible for the stability of the Fermi liquid. In the intermediate-energy region $|\tilde{\xi}| \sim \omega_D$, however, the decay is comparable with the energy and the quasiparticle spectrum looses its meaning. In the high-energy region $|\tilde{\xi}| >> \omega_D$ the decay remains the same in absolute value but again becomes small in comparison with $|\tilde{\xi}|$ and the quasiparticle concept recovers its meaning.

To calculate the phonon self-energy it is convenient first to integrate over frequency in Eq.(3.15) with the following result:

$$\Pi(\mathbf{q},\omega) = \frac{E_p}{2\pi^3} \int d\mathbf{k} \frac{\Theta(\xi_{\mathbf{k}}) - \Theta(\xi_{\mathbf{k}+\mathbf{q}})}{\omega + \xi_{\mathbf{k}} - \xi_{\mathbf{k}+\mathbf{q}} + i0^+ sgn(\xi_{\mathbf{k}+\mathbf{q}})} \tag{3.27}$$

Because of adiabaticity $s << v_F$ and $\omega << v_F q$. Thus one can take $\omega = 0$ in $Re\Pi$ and the first order term in ω in $Im\Pi$:

$$Re\Pi = \lambda h\left(\frac{q}{2k_F}\right) \tag{3.28}$$

$$Im\Pi = \frac{\pi\lambda m}{qk_F}|\omega|\Theta(2k_F - q) \tag{3.29}$$

with $h(x) = 1 + \frac{1-x^2}{2x}\ln\left|\frac{1+x}{1-x}\right|$.

Substituting these expressions into Eq.(3.9) one obtains the phonon Green's function

$$D(\mathbf{q},\omega) = \frac{\omega_{\mathbf{q}}^2}{\omega^2 - \omega_{\mathbf{q}}^2 \left[1 - \lambda h(\frac{q}{2k_F}) + i\frac{\pi\lambda m}{qk_F}|\omega|\Theta(2k_F - q)\right]} \quad (3.30)$$

The pole of D determines a new phonon dispersion $\tilde{\omega}$ and a damping of phonons Γ due to the interaction with electrons:

$$\tilde{\omega}_{\mathbf{q}} = \omega_{\mathbf{q}}\sqrt{1 - \lambda h(\frac{q}{2k_F})} \quad (3.31)$$

$$\Gamma_{\mathbf{q}} = \frac{\pi\lambda m}{2qk_F}\omega_{\mathbf{q}}^2\Theta(2k_F - q) \quad (3.32)$$

The damping Γ is small because of adiabaticity:

$$\Gamma \sim \frac{s}{v_F}\omega << \omega \quad (3.33)$$

However the softening of phonons can be large. At $\lambda > 1/2$ the Migdal approach does not work because long-wave phonons with $q << k_F$ are unstable:

$$\tilde{\omega}_{\mathbf{q}} = \omega_{\mathbf{q}}\sqrt{1 - 2\lambda} \quad (3.34)$$

There is also a more general problem with the application of the Fröhlich hamiltonian to metals. The quadratic term in the ion displacement in the expansion of the pseudopotential Eq.(2.5)

$$v(\mathbf{r} - \mathbf{R}_l) \simeq v(\mathbf{r} - l) - \mathbf{u}_l \cdot \nabla v(\mathbf{r} - l) + \frac{1}{2}u_l^\alpha u_l^\beta \nabla_\alpha \nabla_\beta v(\mathbf{r} - l) \quad (3.35)$$

contributes to the non-Fröhlich electron-lattice interaction. In a simple metal it gives the contribution to the phonon self-energy in the first order practically of the same value as the opposite sign second order contribution from the Fröhlich interaction. This compensation leads to an adiabatically small renormalisation of the acoustic phonon frequency, contrary to the large Migdal softening (Brovman and Kagan (1967)). That makes unacceptable the application of the linear Fröhlich interaction with the acoustic phonons to wide band metals (for review of this problem and of the adiabatic theory of metals see Geilikman (1975)). To apply the linear Fröhlich hamiltonian alone one can include the term quadratic in displacements in the definition of the harmonic phonons. This leads to a bare phonon which is the ion plasmon ($\omega_{\mathbf{q}} = \sqrt{4\pi N e^2/M}$) rather than acoustical mode and the coupling constant with electrons of order of unity. In this case the renormalisation of a bare phonon (plasmon) frequency due to the electron-phonon interaction is of order of the plasmon frequency itself (see Chapter 4). All known high-T_c oxides are doped semiconductors with a narrow electron band and with a well-defined phonon spectrum in the undoped insulating state. This allows us to apply the Fröhlich interaction both to carriers and phonons with the bare phonons from insulating parent compounds.

3.2   Eliashberg extension of the BCS theory

The Migdal approach is valid for a small and intermediate coupling $\lambda < 1/2$ in the normal state. However in a superconductor below T_c macroscopic anomalous averages $< cc >$ and $< c^{\dagger}c^{\dagger} >$ appear because of the Cooper instability of the Fermi liquid versus pairing. Following Eliashberg(1960) one can reformulate the Migdal approach to include these averages in the main diagrams, Fig.3.1. At finite temperature the diagram technique can be formulated for the 'temperature' GF defined by Matsubara(1955) as

$$g(\mathbf{k}, \tau) = -\langle\langle T_\tau c_{\mathbf{k}}(\tau)c_{\mathbf{k}}^{\dagger}\rangle\rangle, \qquad (3.36)$$

with $c_{\mathbf{k}}(\tau) = exp(H\tau)c_{\mathbf{k}}exp(-H\tau)$ and $0 < \tau < 1/T$ a 'thermodynamic' time. The double angular brackets correspond to the quantum as well as statistical average with the Gibbs distribution:

$$\langle\langle...\rangle\rangle = \sum_{\nu} e^{\frac{\Omega - E_\nu}{T}}\langle\nu|...|\nu\rangle \qquad (3.37)$$

where Ω is the thermodynamic potential and $|\nu\rangle$ the eigenstates of $H - \mu N$ with the eigenvalues E_ν. Because the thermodynamic time is restricted by $1/T$ the temperature GF is expanded in the Fourier series:

$$g(\mathbf{k}, \tau) = T\sum_{\omega_n} e^{-i\omega_n\tau}g(\mathbf{k}, \omega_n) \qquad (3.38)$$

with the discrete Matsubara frequencies $\omega_n = \pi T(2n+1)$, $n = 0, \pm 1, \pm 2, \ldots.$ For free electrons:

$$g^{(0)}(\mathbf{k}, \omega_n) = \frac{1}{i\omega_n - \xi_{\mathbf{k}}} \qquad (3.39)$$

and for free phonons:

$$d^{(0)}(\mathbf{q}, \omega_n - \omega_{n'}) = -\frac{\omega_{\mathbf{q}}^2}{[\omega_n - \omega_{n'}]^2 + \omega_{\mathbf{q}}^2} \qquad (3.40)$$

To take into account the Cooper pairing of two electrons with the opposite momentum and spin one can introduce, following Gor'kov (1958) and Nambu (1960) the matrix GF:

$$\hat{g}(\mathbf{k}, \tau) = - \begin{pmatrix} \langle\langle T_\tau c_{\mathbf{k},\uparrow}(\tau) c_{\mathbf{k},\uparrow}^\dagger \rangle\rangle & \langle\langle T_\tau c_{\mathbf{k},\uparrow}(\tau) c_{-\mathbf{k},\downarrow}(\tau)\rangle\rangle \\ \langle\langle T_\tau c_{-\mathbf{k},\downarrow}^\dagger(\tau) c_{\mathbf{k},\uparrow}^\dagger\rangle\rangle & \langle\langle T_\tau c_{-\mathbf{k},\downarrow}^\dagger(\tau) c_{-\mathbf{k},\downarrow}\rangle\rangle \end{pmatrix}. \qquad (3.41)$$

The matrix self-energy

$$\hat{\Sigma}(\mathbf{k}, \omega_n) = \left(\hat{g}^{(0)}(\mathbf{k}, \omega_n)\right)^{-1} - \hat{g}^{-1}(\mathbf{k}, \omega_n) \qquad (3.42)$$

with $\hat{g}^{(0)}(\mathbf{k}, \omega_n) = (i\omega_n \tau_0 - \xi_{\mathbf{k}} \tau_3)^{-1}$. Here $\tau_{0,1,2,3}$ is a set of the Pauli matrices:

$$\tau_0 = \begin{pmatrix} 1 & 0 \\ 0 & 1 \end{pmatrix},$$

$$\tau_1 = \begin{pmatrix} 0 & 1 \\ 1 & 0 \end{pmatrix},$$

$$\tau_2 = \begin{pmatrix} 0 & i \\ i & 0 \end{pmatrix},$$

$$\tau_3 = \begin{pmatrix} 1 & 0 \\ 0 & -1 \end{pmatrix}.$$

The generalized equation for the matrix $\hat{\Sigma}$ is given by the same diagram as in the normal state, Fig.3.1a, with only substitution $\gamma(\mathbf{q})\tau_3$ instead of $\gamma(\mathbf{q})$ and summation over Matsubara frequencies instead of integration:

$$\hat{\Sigma}(\mathbf{k}, \omega_n) = -T \sum_{\omega_{n'}} \int \frac{d\mathbf{q}}{(2\pi)^3} \gamma^2(\mathbf{q}) \omega_{\mathbf{q}} \tau_3 \hat{g}(\mathbf{k} - \mathbf{q}, \omega_{n'}) \tau_3 d(\mathbf{q}, \omega_n - \omega_{n'})$$

$$(3.43)$$

The most important difference of Eq.(3.43) from the normal state Eq.(3.10) is the possibility of obtaining a finite value of the off-diagonal matrix elements (anomalous averages) within the self-consistent solution. In the second $(\hat{g} \to \hat{g}^{(0)})$ or in any finite order of the perturbation theory there are no anomalous averages from Eq.(3.43). That means that within the perturbation theory there is no superconducting phase transition. However if one sums all diagrams of the type Fig.3.1a which means solving Eq.(3.43) selfconsistently, one obtains the finite anomalous averages. To illustrate this we adopt a particular momentum dependence of the interaction constant as in the previous Section and an approximate form of the phonon Green's function:

$$d(\mathbf{q}, \omega_n - \omega_{n'}) \simeq -1 \qquad (3.44)$$

if $|\omega_n - \omega_{n'}| < \omega_D$ and zero otherwise. If there is no current in the system the phase of the order parameter can be chosen to be zero. In this case $\hat{\Sigma}$ is a sum of three Pauli matrices $\tau_{0,1,3}$ with the coefficients $(1 - Z)i\omega_n$, Δ and χ respectively, which are the functions of frequency and momentum:

$$\hat{\Sigma}(\mathbf{k}, \omega_n) = (1 - Z)i\omega_n \tau_0 + \Delta \tau_1 + \chi \tau_3. \qquad (3.45)$$

Thus

$$\hat{g}^{-1}(\mathbf{k}, \omega_n) = Zi\omega_n \tau_0 - \Delta \tau_1 - \tilde{\xi} \tau_3 \qquad (3.46)$$

and

$$\hat{g}(\mathbf{k}, \omega_n) = -\frac{Zi\omega_n + \Delta \tau_1 + \tilde{\xi} \tau_3}{Z^2 \omega_n^2 + \tilde{\xi}^2 + \Delta^2} \qquad (3.47)$$

with $\tilde{\xi} = \xi + \chi$. Substitution of Eq.(3.44,47) in the master equation (3.43) yields

$$(1 - Z)i\omega_n = -\lambda T \int d\tilde{\xi} \sum_{\omega_{n'}} \frac{i\omega_{n'} Z}{Z^2 \omega_{n'}^2 + \tilde{\xi}^2 + \Delta^2} = 0 \qquad (3.48)$$

$$\chi = -\lambda T \int d\tilde{\xi} \sum_{\omega_{n'}} \frac{\tilde{\xi}}{Z^2 \omega_{n'}^2 + \tilde{\xi}^2 + \Delta^2} = 0 \qquad (3.49)$$

$$\Delta = \lambda T \int d\tilde{\xi} \sum_{\omega_{n'}} \frac{\Delta}{Z^2 \omega_{n'}^2 + \tilde{\xi}^2 + \Delta^2} \qquad (3.50)$$

From Eq.(3.48,49) $Z = 1$ and $\chi = 0$. Applying the formula for *tanh* to the sum in Eq.(3.50) we obtain the familiar BCS equation for the order parameter

$$1 = \frac{\lambda}{2} \int \frac{d\xi}{\sqrt{\xi^2 + \Delta^2}} tanh \frac{\sqrt{\xi^2 + \Delta^2}}{2T}, \qquad (3.51)$$

where the integration is restricted by the region $|\xi| < \omega_D$ because of the approximation, Eq.(3.44) for $d(\mathbf{q}, \omega_n - \omega_{n'})$.

There is no direct physical meaning of poles of the temperature GF. To derive the one-particle excitation spectrum one has to calculate the real-time GF determined at finite temperature as

$$G(\mathbf{k}, t) = -i\langle\langle T_t c_{\mathbf{k}}(t) c_{\mathbf{k}}^\dagger \rangle\rangle, \qquad (3.52)$$

with the real time t. One can use the retarded G^R and advanced G^A Green functions:

$$G^R(\mathbf{k}, t) = -i\Theta(t)\langle\langle [c_{\mathbf{k}}(t) c_{\mathbf{k}}^\dagger]\rangle\rangle \qquad (3.53)$$

$$G^A(\mathbf{k}, t) = i\Theta(-t)\langle\langle [c_{\mathbf{k}}(t) c_{\mathbf{k}}^\dagger]\rangle\rangle \qquad (3.54)$$

where [...] is an anticommutator. They are analytical in the upper or lower half-plane of ω correspondingly. There is a simple connection between the Fourier components of G and $G^{R,A}$

$$G^{R,A}(\mathbf{k}, \omega) = ReG(\mathbf{k}, \omega) \pm icoth(\frac{\omega}{2T})ImG(\mathbf{k}, \omega) \qquad (3.55)$$

from one side and between those of $G^{R,A}$ and g from another

$$G^R(\mathbf{k}, i\omega_n) = g(\mathbf{k}, \omega_n) \qquad (3.56)$$

for $\omega_n > 0$ and

$$g(\mathbf{k}, -\omega_n) = g^*(\mathbf{k}, \omega_n). \qquad (3.57)$$

In our case the temperature GF

$$g(\mathbf{k}, \omega) = \frac{u_{\mathbf{k}}^2}{i\omega_n - \epsilon_{\mathbf{k}}} + \frac{v_{\mathbf{k}}^2}{i\omega_n + \epsilon_{\mathbf{k}}}. \qquad (3.58)$$

where $u_{\mathbf{k}}^2, v_{\mathbf{k}}^2 = (\epsilon_{\mathbf{k}} \pm \xi_{\mathbf{k}})/2\epsilon_{\mathbf{k}}$ and $\epsilon_{\mathbf{k}} = \sqrt{\xi_{\mathbf{k}}^2 + \Delta^2}$. The analytical continuation of this expression to the upper half-plane yields

$$G^R(\mathbf{k}, \omega) = \frac{u_{\mathbf{k}}^2}{\omega - \epsilon_{\mathbf{k}} + i0^+} + \frac{v_{\mathbf{k}}^2}{\omega + \epsilon_{\mathbf{k}} + i0^+}. \qquad (3.59)$$

and with Eq.(1.55) one obtains

$$
\begin{aligned}
G(\mathbf{k},\omega) \;=\; & Re\left(\frac{u_{\mathbf{k}}^2}{\omega - \epsilon_{\mathbf{k}}} + \frac{v_{\mathbf{k}}^2}{\omega + \epsilon_{\mathbf{k}}}\right) \\
& - i\pi \tanh(\frac{\omega}{2T})\left(u_{\mathbf{k}}^2 \delta(\omega - \epsilon_{\mathbf{k}}) + v_{\mathbf{k}}^2 \delta(\omega + \epsilon_{\mathbf{k}})\right) \quad (3.60)
\end{aligned}
$$

For $T = 0$, $\tanh(\frac{\omega}{2T}) = sgn(\omega)$ and

$$
G(\mathbf{k},\omega) = \frac{u_{\mathbf{k}}^2}{\omega - \epsilon_{\mathbf{k}} + i0+} + \frac{v_{\mathbf{k}}^2}{\omega + \epsilon_{\mathbf{k}} - i0+}. \quad (3.61)
$$

The poles of GF are the same as the BCS quasiparticles:

$$
\epsilon_{\mathbf{k}} = \sqrt{\xi_{\mathbf{k}}^2 + \Delta^2(0)}. \quad (3.62)
$$

Thus the Migdal-Eliashberg theory reproduces the BCS results if a similar approximation for the attraction between electrons is made. Within a more general consideration the master equation (3.43) takes properly into account the phonon spectrum, retardation and realistic matrix element of the electron-phonon interaction (Scalapino (1969)). In particular it is useful in a study of the effect of the Coulomb repulsion on the pairing. There is no adiabatic parameter for this interaction. Nevertheless in a qualitative analysis one can adopt the same contribution to the electron self-energy from the Coulomb interaction as from phonons replacing $\gamma^2(\mathbf{q})\omega_{\mathbf{q}}d(\mathbf{q},\omega_n - \omega_{n'})$ in Eq.(3.43) for the Fourier component of the Coulomb potential $4\pi e^2/\epsilon q^2$ with ϵ a dielectric constant of a host material. The Coulomb interaction is nonretarded for the frequencies less or compared with the Fermi energy, so the equation for the order parameter becomes

$$
\Delta(\omega_n) = T \int d\xi \sum_{\omega_{n'}} K(\omega_n - \omega_{n'}) \frac{\Delta(\omega_{n'})}{\omega_{n'}^2 + \xi^2 + \Delta^2(\omega_{n'})} \quad (3.63)
$$

where the kernel K is given by

$$
K(\omega_n - \omega_{n'}) = \lambda\Theta(\omega_D - |\omega_n - \omega_{n'}|) - \mu_c\Theta(\mu - |\omega_n - \omega_{n'}|) \quad (3.64)
$$

with μ_c the product of the Fourier component of the Coulomb potential and the normal density of states at the Fermi level. At

$T = T_c$ one can neglect the second power of the order parameter in Eq.(3.63) and integrating over ξ obtain

$$\Delta(\omega_n) = \pi T_c \sum_{\omega_{n'}} K(\omega_n - \omega_{n'}) \frac{\Delta(\omega_{n'})}{|\omega_{n'}|} \qquad (3.65)$$

To solve this equation we adopt the BCS-like parametrization of the kernel

$$\begin{aligned} K(\omega_n - \omega_{n'}) &\simeq \lambda\Theta(2\omega_D - |\omega_n|)\Theta(2\omega_D - |\omega_{n'}|) \\ &- \mu_c\Theta(2\mu - |\omega_n|)\Theta(2\mu - |\omega_{n'}|) \end{aligned} \qquad (3.66)$$

and replace the summation by the integration

$$\pi T_c \sum \to \int_{\pi T_c}^{\infty} d\omega \qquad (3.67)$$

because $T_c << \omega_D, \mu$. The solution can be found in the form

$$\Delta(\omega) = \Delta_1\Theta(2\omega_D - |\omega|) + \Delta_2\Theta(2\mu - |\omega|)\Theta(|\omega| - 2\omega_D) \qquad (3.68)$$

with constant but different values of the order parameter below (Δ_1) and above (Δ_2) the cut-off energy $2\omega_D$. Substitution of Eq.(3.66,68) into Eq.(3.65) yields for $\Delta_{1,2}$

$$\Delta_1\left[1 - (\lambda - \mu_c)\ln\frac{2\omega_D}{\pi T_c}\right] + \Delta_2\mu_c\ln\frac{\mu}{\omega_D} = 0 \qquad (3.69)$$

$$\Delta_1\mu_c\ln\frac{2\omega_D}{\pi T_c} + \Delta_2\left[1 + \mu_c\ln\frac{\mu}{\omega_D}\right] = 0 \qquad (3.70)$$

The condition of the existence of a nontrivial solution of these coupled equations gives T_c

$$T_c = \frac{2\omega_D}{\pi}exp\left(-\frac{1}{\lambda - \mu_c^*}\right) \qquad (3.71)$$

where

$$\mu_c^* = \frac{\mu_c}{1 + \mu_c\ln(\mu/\omega_D)} \qquad (3.72)$$

is the Coulomb pseudopotential (Tolmachev (1958), Morel and Anderson (1962)). This is a remarkable result. It shows that even a large Coulomb repulsion $\mu_c > \lambda$ does not destroy Cooper pairs because its contribution is suppressed down to the value

$\sim 1/\ln\frac{\mu}{\omega_D} \ll 1$. The retarded attraction mediated by phonons acts well after two electrons meet each other. This time delay is sufficient for two electrons to be separated by the relative distance at which the Coulomb repulsion is small.

For metals and their alloys the empirical McMillan (1968) formula for T_c is adopted:

$$T_c = \frac{\omega_D}{1.45}exp\left(-\frac{1.04(1+\lambda)}{\lambda-\mu_c^*(1+0.62\lambda)}\right) \tag{3.73}$$

which works well for low-T_c materials even if the estimated λ is large (> 1) as in Pb. However already in materials with a moderate $T_c \sim 20K$ as in $A-15$ compounds (Nb_3Sn, V_3Si) the discrepancy in the values of λ estimated with Eq.(3.73) and with the direct band-structure calculations exceeds by several times the limit allowed by the experimental and computation accuracy (Klein *et al.* (1978)) and thereby the credibility of the canonical BCS approach to these materials is low. Small polarons were proposed for $A-15$ compounds to explain this inconsistency (Alexandrov and Elesin (1983)).

The Migdal-Eliashberg theory is based on the assumption that the Fermi liquid is stable, therefore the condition $\mu \gg \omega_D$ is satisfied. In original papers Migdal (1958) and Eliashberg (1960) restricted the region of the applicability of their approach by the value of the coupling $\lambda < 1$. In the following chapters we show that the proper extension of the BCS theory to the strong coupling region $\lambda > 1$ inevitably involves the small polaron formation. The critical value of λ at which polaron forms depends on the type of the electron-phonon interaction. In simple wide band metals like Pb the crystal field is screened, and the phonon assisted intersite hopping dominates in the electron-phonon interaction. This hopping can destroy polaron. That is why Pb with $\lambda > 1$ shows no sign of polarons. However small polarons are formed already at $\lambda < 1$ in narrow band materials.

Chapter 4

Strongly coupled electrons and phonons

4.1 Breakdown of the Migdal approach

In the adiabatic Migdal description of coupled electrons and phonons the instability appears at the bare coupling constant $\lambda \simeq 1/2$: the phonon frequency renormalised by the interaction $\tilde{\omega} \simeq \omega\sqrt{1-2\lambda}$ (Chapter 3). In a more general sense this is a boson vacuum instability driven by the strong interaction of fermionic and bosonic fields.

The instability of the Fermi-liquid or BCS liquid (below T_c) strongly coupled with phonons can be also seen in the electron self-energy. The electron self-energy Σ in the Migdal approximation contains two contributions, Σ_M, Fig.3.1a and Σ_μ, Fig.3.1d. $\Sigma_M \simeq \lambda\omega$ and thus remains small compared with the bandwidth $2D \simeq N(0)^{-1}$ in the relevant region of the coupling ($\lambda < D/\omega$), which guarantees the self-consistency of the approach. On the other hand for optical or molecular phonons $\Sigma_\mu \simeq D\lambda n$ turns out to be comparable or larger than the Fermi energy already at $\lambda \sim 1$ for any filling of the band (n is the electron density per cell). As a rule this diagram which is momentum and frequency independent, is included in the definition of the chemical potential μ. However this is justified only for a weak-coupling regime. For a strong coupling Σ_μ leads to an instability. To show this let us consider a one-dimensional chain in the tight-binding approximation with the nearest-neighbor hopping integral $D/2$. The renormalised chemical potential is given by

$$\mu = D sin \left(\frac{\pi(n-1)}{2} \right) - 2D\lambda n. \tag{4.1}$$

The system is stable if $d\mu/dn$ is positive, which yields the following region of the stability of the Migdal solution:

$$\lambda < \frac{\pi}{2} cos \left(\frac{\pi(n-1)}{2} \right) \tag{4.2}$$

For two and three-dimensional lattices the numerical coefficient is different , but the critical value of λ remains of the order of unity. This consideration shows that the extension of the Migdal approximation to the strong-coupling region $\lambda > 1$ is unacceptable. For a filling different from $1/2$ the critical value of λ above which the Migdal approximation is violated turns out to be even less. In the following we show that depending on the value of the Coulomb repulsion a many electron system strongly coupled with any bosonic field is a polaronic Fermi liquid or a bipolaronic Bose liquid.

4.2 Small polarons and bipolarons

The coupling constant λ is the ratio of the characteristic interaction energy E_p of carriers with a bosonic field, for instance of phonons, which is responsible for the coupling to their kinetic energy $E_F \sim D$, $\lambda = E_p/E_F$. At the point $\lambda \simeq 1$ the characteristic potential energy due to the lattice deformation exceeds the kinetic energy. This is a condition for a small polaron formation which was known for a long time as a solution for a single electron on a lattice coupled with lattice vibrations.

There is some confusion in the literature, particularly about the use of the terms 'large' and 'small' polaron and bipolaron. This section attempts to set out what we actually know about these entities.

The concept of polarons was first introduced by Landau (1933). If an electron is placed into the conduction band of an ionic crystal the force on another electron at a distance r from it would be

$e^2/\epsilon_0 r^2$. But if the ions did not move the force would be $e^2/\epsilon r^2$, where ϵ_0 and ϵ are the static and high-frequency dielectric constant respectively. Thus an electron is acted upon by a potential energy

$$- \frac{e^2(\epsilon^{-1} - \epsilon_0^{-1})}{r} \tag{4.3}$$

Since this is a Coulomb potential, localised states must exist. The electron is 'trapped by digging its own hole'. Mott and Gurney (1940) argued that the trapped electron must be mobile but heavy, and found it remarkable that no such entity had been observed. The concept was treated in much greater detail by Fröhlich (1954), Yamashita and Kurosawa (1958), Sewell (1958), Holstein (1959), Toyozawa (1961), Friedman (1964), Eagles (1966), Holstein and Friedman (1968), Austin and Mott (1969), Emin and Holstein (1969), Emin (1970), the Russian school: Pekar (1951), Tjablikov (1952), Rashba (1957), Klinger (1961), Lang and Firsov (1962) and later by many others. In Holstein's treatment an electron is trapped by the self induced deformation of two atomic molecules. A Franck-Condon model is used to calculate their quasistatic displacement and the frequency with which the polaron can move to a neighboring molecule. This (in the so-called adiabatic approximation) involves the excitation of both the occupied and empty molecules to the same energy, so that the electron can tunnel backwards and forwards between them. This excitation involves an activation energy $E_p/2$, and the excitation can occur through the action of temperature or zero-point motion. In the former case the polaron moves by thermally activated hopping, with a diffusion coefficient

$$\omega a^2 exp\left(-\frac{E_p}{2T}\right) \tag{4.4}$$

where a is the distance between molecules. In the latter case, for $T < \omega/2$ the motion is coherent, the polaron behaves like a heavy particle with the mass

$$m^* \sim exp\left(\frac{E_p}{\omega}\right) \tag{4.5}$$

and with a mean free path determined by the phonon and impurity scattering. Here ω is the characteristic phonon frequency. For ionic crystals we may write

$$E_p = \frac{e^2(\epsilon^{-1} - \epsilon_0^{-1})}{r_p} \tag{4.6}$$

where r_p is the radius of the volume within the wave function trapped by the lattice deformation (polaron radius). This can be greater than or comparable with the lattice constant a. This formula applies to ionic lattices.

It is of course possible that the process described only leads to a small phase shift in the wave function at each hop if $E_p \ll 1/ma^2$, where m is a rigid band mass. The polaron is than called 'large', and can be described as a free particle moving in an elastic continuum. The most sophisticated treatment of the large or 'continuum' polaron is due to Feynman and co-workers (1962) following that given by Pekar (1951) and Fröhlich (1954). This treatment leads to a mass enhancement and several maxima in the optical absorption, but not to a hopping conduction or to a narrow polaron band. It is in this sense that we use the term 'large polaron'. Some authors used the term 'large' to describe only situations where $(\epsilon^{-1} - \epsilon_0^{-1})$ is considerably less than unity. But even this is by no means the case of large polarons if the bare electron band is sufficiently narrow.

For non-ionic materials (elements), large polarons in our sense do not exist. If we take the Holstein model with the phonon coordinate u (molecular deformation) which lowers the electron energy by $u\omega\sqrt{2ME_p}$, then the deformation region of the size r_p with one localised electron in it costs the energy

$$E(u, r_p) \simeq \frac{1}{2mr_p^2} - u\omega\sqrt{2ME_p} + \frac{M\omega^2 u^2}{2}(\frac{r_p}{a})^3 \tag{4.7}$$

Minimising it with respect to the displacement one obtains

$$u_0 = \sqrt{\frac{2E_p}{M\omega^2}}(\frac{a}{r_p})^3 \tag{4.8}$$

and the minimum

$$E(u_0, r_p) = \frac{1}{2mr_p^2} - E_p(\frac{a}{r_p})^3, \tag{4.9}$$

The energy has a minimum for $r_p = 0$; therefore the electron collapses into a point at any value of the polaron binding energy E_p. In fact r_p is restricted from below by the lattice constant a. Therefore there is a critical value of the coupling constant $\lambda = 2ma^2 E_p$:

$$\lambda_c = 1. \tag{4.10}$$

No polarons are formed in the weak coupling region $\lambda < \lambda_c$ ($r_p = \infty$), and only small polarons ($r_p = a$) exist in the strong-coupling regime $\lambda > \lambda_c$.

The concept of a small polaron was first applied to an electron or hole in a non-metal (e.g. in alkali halides). For the new superconductors our model is of course that of a degenerate gas of small polarons or a gas of bipolarons which is also degenerate below T_c and nondegenerate above. Mott and Davis (1979) discussed the first possibility. Mott (1973) has also discussed overcrowding, namely that the concentration of polarons must be sufficiently low, about one per several atomic sites to avoid the overlapping of their deformation fields. Materials containing a degenerate gas of small polarons were thought to be $Na_x V_2 O_5$ and $K_{0.3} MnO_3$. The latter is diamagnetic at low temperatures, suggesting bipolarons, but probably too heavy to move: otherwise superconductivity would be expected. Superconducting $BaBiPbO$ ($T_c \simeq 13K$) and the heavy fermion superconductors were supposed to be (bi)polaronic (Alexandrov *et al.* (1986a,b)).

The concept of a small on-site *localised bipolaron* was introduced by Anderson (1975) and by Street and Mott (1975). It was applied to glassy semiconductors (chalcogenide glasses) to explain their magnetic and electric properties (Anderson (1975), Street and Mott (1975), Mott, Davis and Street (1975), Klinger and Karpov (1982) and others). Small bipolarons different from those were identified by Lakkis *et al.* (1976) in $Ti_4 O_7$ and $Ti_{4-x} V_x O_7$ and Chakraverty *et al.* (1978) in $Na_x V_2 O_5$ using the magnetic susceptibility, heat capacity, resistivity and other measurements. These are intersite small bipolarons, which are bound states of two Ti^{+3} ions stabilised by a large lattice distortion. At low temperatures they form a charged ordered state and are immobile. Nevertheless at a higher temperature, depending on the level of *vanadium* doping, a dynamical disorder of $Ti^{+3} - Ti^{+3}$ pairs was observed.

Chakraverty (1981) and Chakraverty *et al.* (1987) discussed bipolarons in connection with superconductivity, but considered that bipolarons would be immobile and therefore inactive as superconducting carriers.

The introduction of small *mobile* bipolarons and small polarons into the theory of superconductivity dates from the work of Alexandrov and Ranninger (1981a,b) and Alexandrov (1983) respectively and is discussed in Chapter 5.

Mobile polarons and bipolarons in various materials were observed by Salje and co-workers. They investigated polaron formation in WO_{3-x} and NbO_{3-x} in much detail. In the case of WO_{3-x} Gehlig and Salje (1983) found a changeover from a regime of hopping conductivity with the constant activation energy to a regime with much smaller and zero activation energy, thus expecting polaron band conduction below $130K$. Bipolarons were identified by the absence of a magnetic moment: from the maximum value of x for which they are formed the size of the bipolaron is found to be about 10Å (Ruscher, Salje and Hussain(1988)).

4.3   Exact solution for $\lambda \to \infty$

Because of the selftrapping the Wannier (site) representation is more convenient for the bare Fröhlich hamiltonian in the strong coupling regime $\lambda > 1$:

$$c_i = \frac{1}{\sqrt{N}} \sum_{\mathbf{k}} e^{i\mathbf{k}\cdot\mathbf{m}} c_{\mathbf{k},s} \qquad (4.11)$$

where $i = (\mathbf{m}, s)$ includes both site $\mathbf{m}$ and spin s. In the site representation the sum of the kinetic energy, electron-phonon and the Coulomb interactions and phonon energy is given by

$$
\begin{aligned}
H = & \sum_{i,j}(T(\mathbf{m} - \mathbf{n})\delta_{s,s'} - \mu\delta_{i,j})c_i^\dagger c_j + \sum_{q,i}\omega_q\hat{n}_i\left(u_i(\mathbf{q})d_{\mathbf{q}} + h.c.\right) \\
& + \frac{1}{2}\sum_{i,j}V_c(\mathbf{m} - \mathbf{n})c_i^\dagger c_j^\dagger c_j c_i + \sum_q \omega_q d_{\mathbf{q}}^\dagger d_{\mathbf{q}}
\end{aligned}
\qquad (4.12)
$$

with the bare hopping integral

$$T(\mathbf{m}) = \frac{1}{N}\sum_{\mathbf{k}} E_{\mathbf{k}} e^{i\mathbf{k}\cdot\mathbf{m}}, \qquad (4.13)$$

the matrix element of the electron-phonon interaction

$$u_i(\mathbf{q}) = \frac{1}{\sqrt{2N}}\gamma(\mathbf{q})e^{i\mathbf{q}\cdot\mathbf{m}} \tag{4.14}$$

and the Coulomb interaction

$$V_c(\mathbf{m}) = \frac{1}{N}\sum_{\mathbf{q}}\frac{4\pi e^2}{\epsilon q^2}e^{i\mathbf{q}\cdot\mathbf{m}} \tag{4.15}$$

Here $i = (\mathbf{m}, s)$, $j = (\mathbf{n}, s')$ and $\hat{n}_i = c_i^\dagger c_i$. The dielectric constant ϵ in the definition of the Coulomb potential takes into account the Coulomb interaction with the dielectric matrix, which is accompanied by the polarization of inner electrons of the ions and cannot be included in the crystal field potential $V(\mathbf{r})$, Eq.(2.4).

Strictly speaking the separation of electrons into 'carriers' and 'inner' electrons is a well defined only for semiconductors. In this case the parent dielectric compound exists with well defined bare phonons $\omega_{\mathbf{q}}$ and the electronic band structure $E_{\mathbf{k}}$. For metals the hamiltonian Eq.(4.12) should be treated at best as the bare one, in which all matrix elements and phonon frequencies have no direct physical meaning (see also Section 3.1). In particular taking the electron-phonon matrix element depending only on the momentum transfer one neglects the contribution to the interaction, containing the overlap of different site orbitals, which is a good approximation only for narrow bands, whose bandwidth $2D$ is less than the characteristic value of the crystal field. In simple metals with wide bare bands the hopping term dominates in the electron-phonon interaction making the formation of small polarons more difficult and shifting the critical value of λ towards higher values.

As far as $\lambda > 1$ the kinetic energy remains smaller than the interaction energy and a self-consistent treatment of a many electron system strongly coupled with phonons is possible with the '$1/\lambda$' expansion technique (Alexandrov (1992b)). This possibility results from the fact, known for a long time, that there is an exact solution for a single electron in the strong-coupling limit $\lambda \to \infty$. Following Lang and Firsov (1962) one can apply the canonical transformation e^S to diagonalise the hamiltonian. The diagonalisation is exact if $T(\mathbf{m}) = 0$ (or $\lambda = \infty$):

$$\tilde{H} = e^S H e^{-S}, \tag{4.16}$$

where

$$S = \sum_{q,i} \hat{n}_i \left(u_i(\mathbf{q})d_{\mathbf{q}} - h.c. \right) \tag{4.17}$$

The electron operator transforms as

$$\tilde{c}_i = c_i exp \left(\sum_q u_i(\mathbf{q})d_{\mathbf{q}} - h.c. \right) \tag{4.18}$$

and the phonon one as:

$$\tilde{d}_{\mathbf{q}} = d_{\mathbf{q}} - \sum_i \hat{n}_i u_i(\mathbf{q}) \tag{4.19}$$

From Eq.(4.19) it follows that the Lang-Firsov canonical transformation shifts ions to new equilibrium positions. In a more general sense it changes the boson vacuum. As a result:

$$\tilde{H} = \sum_{i,j} (\hat{\sigma}_{ij} - \mu\delta_{i,j})c_i^{\dagger}c_j - E_p \sum_i \hat{n}_i + \frac{1}{2} \sum_{i,j} v_{ij}c_i^{\dagger}c_j^{\dagger}c_jc_i + \sum_q \omega_q d_q^{\dagger}d_q \tag{4.20}$$

where

$$\hat{\sigma}_{ij} = T(\mathbf{m} - \mathbf{n})\delta_{s.s'} exp \left(\sum_{bfq,i} [u_i(\mathbf{q}) - u_j(\mathbf{q})]d_{\mathbf{q}} - h.c. \right) \tag{4.21}$$

is the new hopping integral depending on the phonon variables,

$$v_{ij} = V_c(\mathbf{m} - \mathbf{n}) - 2 \sum_q \omega_q \left(u_i(\mathbf{q})u_j^*(\mathbf{q}) \right) \tag{4.22}$$

is the polaron-polaron interaction comprising the direct Coulomb repulsion and the attraction via a *nonretarded* lattice deformation (second term of Eq.(4.22)). In a very strong coupling limit $\lambda \to \infty$ one can neglect the hopping term of the transformed Hamiltonian. The rest has analytically determined eigenstates and eigenvalues. The eigenstates $|\check{N}\rangle = |n_i, n_{\mathbf{q}}\rangle$ are classified with the polaron $n_{m,s}$ and phonon $n_{\mathbf{q}}$ occupation numbers and the energy levels are:

$$E = (T(0) - E_p - \mu) \sum_i n_i + \frac{1}{2} \sum_{i,j} v_{ij}n_in_j + \sum_q \omega_q n_q \tag{4.23}$$

where $n_i = 0, 1$ and $n_{\mathbf{q}} = 0, 1, 2, 3, ...\infty$, and the interaction term does not include the on-site interaction $i = j$ for parallel spins because of the Pauli principle.

Thus we conclude that the hamiltonian Eq.(4.12) in zero order of the hopping describes localised polarons and independent phonons which are vibrations of ions relative to new equilibrium positions depending on the polaron occupation numbers. The phonon frequencies remain unchanged in this limit. The middle of the electronic band $T(0)$ falls down by E_p as a result of a potential well produced by the lattice deformation due to the selftrapping (see Fig.4.1).

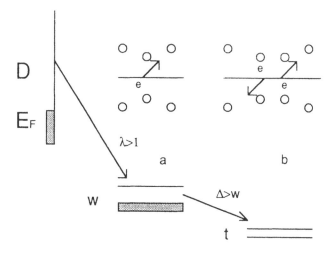

Fig.4.1. Polaron collapse of the electron band (*a*); bipolaron band (*b*).

4.4   Polaron band and self-energy

With the finite hopping term polarons tunnel in a narrow band because of the degeneracy of the zero order hamiltonian relative the site position of a single polaron in a regular lattice. To show this selfconsistently one can apply perturbation theory using $1/\lambda$ as a small parameter. Because of the degeneracy terms of the first order in $T(\mathbf{m})$ should be included in a zero order hamiltonian H_0:

$$H_0 = \sum_{i,j}(\sigma(\mathbf{m}-\mathbf{n}) - \mu\delta_{i,j})c_i^\dagger c_j + \sum_{\mathbf{q}}\omega_{\mathbf{q}}d_{\mathbf{q}}^\dagger d_{\mathbf{q}} \qquad (4.24)$$

where

$$\sigma(\mathbf{m}-\mathbf{n}) = \langle\langle\hat{\sigma}_{ij}\rangle\rangle = T(\mathbf{m}-\mathbf{n})\delta_{s,s'}exp[-g^2(\mathbf{m}-\mathbf{n})] \qquad (4.25)$$

is the hopping integral averaged with the phonon equilibrium distribution,

$$g^2(\mathbf{m}) = \frac{1}{2N}\sum_{\mathbf{q}}\gamma^2(\mathbf{q})coth\left(\frac{\omega_{\mathbf{q}}}{2T}\right)[1 - cos(\mathbf{q}\cdot\mathbf{m})] \qquad (4.26)$$

and T is the temperature. The renormalised position of the middle of the electron band is chosen to be zero, so that $T(0) - E_p = 0$.

The interaction term in the transformed hamiltonian $\tilde{H} = H_0 + H_{int}$ includes the residual interaction of polarons with phonons H_{p-ph} and the polaron-polaron one H_{p-p}:

$$H_{int} = H_{p-ph} + H_{p-p} \qquad (4.27)$$

where

$$H_{p-ph} = \sum_{i,j}[\hat{\sigma}_{ij} - \sigma(\mathbf{m}-\mathbf{n})]c_i^\dagger c_j \qquad (4.28)$$

and

$$H_{p-p} = \frac{1}{2}\sum_{i,j}v_{ij}c_i^\dagger c_j^\dagger c_j c_i \qquad (4.29)$$

The polaron-phonon interaction leads to the polaron bandwidth and phonon frequency renormalisations and to the scattering of polarons. The polaron-polaron correlations are responsible for the scattering and in case of the attraction for the bipolaron formation. If the temperature is well above the temperature for the formation of bipolarons (see below) one can treat H_{p-ph} with the $1/\lambda$ perturbation expansion and H_{p-p} with the canonical random phase approximation (RPA). We consider first the effect of H_{p-ph} on the polaron self-energy Σ_p.

Because of the translation symmetry H_0 is diagonal in the Bloch-representation Eq.(4.11):

$$H_0 = \sum_{\mathbf{k},s}(\epsilon_{\mathbf{k}} - \mu)c_{\mathbf{k},s}^\dagger c_{\mathbf{k},s} + \sum_{\mathbf{q}}\omega_{\mathbf{q}}d_{\mathbf{q}}^\dagger d_{\mathbf{q}} \qquad (4.30)$$

with a polaronic band

$$\epsilon_{\mathbf{k}} = \sum_{\mathbf{m}} \sigma(\mathbf{m}) e^{i\mathbf{k}\cdot\mathbf{m}} \tag{4.31}$$

From Eq.(4.31) we conclude that the polaronic band has the band-width $2w$ exponentially reduced compared with the bare electronic bandwidth:

$$w = De^{-g^2}, \tag{4.32}$$

where g is determined from Eq.(4.26) for the nearest neighbour hopping $|\mathbf{m}| = a$. An increase of the effective mass $m^* = 1/wa^2$ is due to the phonon cloud surrounding a small polaron.

To derive the polaron self-energy one can solve perturbatively the equation of motion for the polaron temperature Green's function defined in the Wannier representation as

$$\tilde{g}_{ij}(\tau) = -\langle\langle T_\tau c_i(\tau) c_j^\dagger \rangle\rangle, \tag{4.33}$$

with $c_i(\tau) = e^{\tilde{H}\tau} c_i e^{-\tilde{H}\tau}$. The double angular brackets correspond now to the quantum as well as statistical average with the eigenstates of the transformed hamiltonian. Differentiating Eq.(4.33) we obtain

$$\frac{d\tilde{g}_{ij}(\tau)}{d\tau} - \mu\tilde{g}_{ij}(\tau) = -\delta(\tau)\delta_{i,j} + \sum_{i'}\langle\langle T_\tau \hat{\sigma}_{ii'}(\tau) c_{i'}(\tau) c_j^\dagger \rangle\rangle \tag{4.34}$$

where $\hat{\sigma}_{ii'}(\tau) = e^{\tilde{H}\tau} \hat{\sigma}_{ii'} e^{-\tilde{H}\tau}$.

For the Fourier component we have:

$$\hat{\tilde{g}}^{-1}(\omega_n) = \left(\hat{\tilde{g}}^{(0)}(\omega_n)\right)^{-1} - \hat{\Sigma}(\omega_n), \tag{4.35}$$

where $\hat{\tilde{g}}$ and $\hat{\Sigma}$ are matrices with respect to site indices i, j. The free polaron GF is determined by

$$\left(\tilde{g}^{(0)}(\omega_n)\right)^{-1}_{ij} = (i\omega_n + \mu)\delta_{i,j} - \sigma(\mathbf{m} - \mathbf{n}) \tag{4.36}$$

and the polaron self-energy by

$$\Sigma_{ij}(\omega_n) = \sum_{i',j'} \Gamma^{i'j'}_{ii'}(\omega_n) \hat{\tilde{g}}^{-1}_{j'j}(\omega_n) \tag{4.37}$$

where $\hat{\Gamma}(\omega_n)$ is the Fourier component of the polaron-phonon correlation function

$$\Gamma_{ii'}^{jj'}(\tau) = -\langle\langle T_\tau[\hat{\sigma}_{ii'}(\tau) - \sigma(\mathbf{m} - \mathbf{m}')]c_j(\tau)c_{j'}^\dagger\rangle\rangle. \qquad (4.38)$$

This correlation function comprises the second and higher order terms in the hopping integral, as follows from the corresponding equation of motion:

$$\frac{d\Gamma_{ii'}^{jj'}(\tau)}{d\tau} - \mu\Gamma_{ii'}^{jj'}(\tau) = \sum_{i''}\langle\langle T_\tau[\hat{\sigma}_{ii'}(\tau) - \sigma(\mathbf{m} - \mathbf{m}')]c_j(\tau)c_{i''}^\dagger\hat{\sigma}_{i''j'}\rangle\rangle.$$
$$(4.39)$$

To obtain Eq.(4.39) one should differentiate the function $\Gamma_{ii'}^{jj'}(-\tau) = -\langle\langle T_\tau[\hat{\sigma}_{ii'} - \sigma(\mathbf{m} - \mathbf{m}')]c_j c_{j'}^\dagger(\tau)\rangle\rangle$. In the second order of the perturbation theory one can replace $\tilde{H}$ in Eq.(4.39) for H_0 to obtain

$$\Sigma_{ij}(\omega_n) = T\sum_{\omega_{n'}}\sum_{i',i'',j'}\tilde{\Phi}_{ii'}^{i''j'}(\omega_n-\omega_{n'})\tilde{g}_{i'i''}^{(0)}(\omega_{n'})\frac{(i\omega_n + \mu)\delta_{j',j} - \sigma(\mathbf{n} - \mathbf{n}')}{i\omega_n + \mu}$$
$$(4.40)$$

where

$$\tilde{\Phi}_{ii'}^{jj'}(\Omega_n) = \Phi_{ii'}^{jj'}(\Omega_n) - \frac{1}{T}\sigma(\mathbf{m} - \mathbf{m}')\sigma(\mathbf{n} - \mathbf{n}') \qquad (4.41)$$

and $\Phi_{ii'}^{jj'}(\Omega_n)$ is a Fourier component of the multiphonon correlator

$$\Phi_{ii'}^{jj'}(\tau) = \langle\langle T_\tau\hat{\sigma}_{ii'}(\tau)\hat{\sigma}_{jj'}\rangle\rangle \qquad (4.42)$$

with $\tilde{H} = H_0$. Direct calculations yield

$$\Phi_{ii'}^{jj'}(\tau) = \sigma(\mathbf{a})\sigma(\mathbf{b})exp\left(\frac{1}{N}\sum_{\mathbf{q}}\gamma^2(\mathbf{q})f_\mathbf{q}\frac{cosh\left[\omega_\mathbf{q}(\frac{1}{2T} - |\tau|)\right]}{sinh\frac{\omega_\mathbf{q}}{2T}}\right)$$
$$(4.43)$$

where $2f_\mathbf{q} = cos(\mathbf{q}\cdot[\mathbf{c} - \mathbf{a}]) + cos(\mathbf{q}\cdot[\mathbf{c} + \mathbf{b}]) - cos(\mathbf{q}\cdot\mathbf{c}) - cos(\mathbf{q}\cdot[\mathbf{c} - \mathbf{a} + \mathbf{b}])$ with $\mathbf{a} = \mathbf{m} - \mathbf{m}'$, $\mathbf{b} = \mathbf{n} - \mathbf{n}'$ and $\mathbf{c} = \mathbf{n}' - \mathbf{m}'$. To simplify Eq.(4.40) further one can neglect all terms, containing a small exponent $e^{-g^2} \ll 1$. Omitting small terms the self-energy is diagonal, $\Sigma_{ij} \simeq \Sigma\delta_{i,j}$ where

$$\Sigma = T\sum_{\omega_{n'}}\sum_{\mathbf{a}}\frac{\Phi_\mathbf{a}(\omega_n - \omega_{n'})}{i\omega_{n'} + \mu} \qquad (4.44)$$

and

$$\Phi_{\mathbf{a}}(\Omega_n) = \frac{1}{2} \int_{-1/T}^{1/T} d\tau\, e^{i\Omega_n \tau} \Phi_{ii'}^{i'i}(\tau) \qquad (4.45)$$

with $\Omega_n = 2\pi nT$; $n = 0, \pm 1, \pm 2, ...$ One can calculate this Fourier component in case of dispersionless phonons $\omega_{\mathbf{q}} = \omega_0$. The main frequency independent contribution is

$$\Phi_{\mathbf{a}}(\Omega_n) \simeq \frac{T^2(\mathbf{a})}{g^2(\mathbf{a})\omega_0} \qquad (4.46)$$

Summation over frequencies in Eq.(4.44) yields

$$T \sum_{\omega_{n'}} \frac{1}{i\omega_{n'} + \mu} = \frac{1}{2} tanh\frac{\mu}{2T} \qquad (4.47)$$

The chemical potential is determined via the atomic density of carriers n

$$\frac{2}{N} \sum_{\mathbf{k}} n_{\mathbf{k}} = n \qquad (4.48)$$

with the Fermi-Dirac distribution function

$$n_{\mathbf{k}} = \frac{1}{exp[(\epsilon_{\mathbf{k}} - \mu)/T] + 1} \simeq \frac{1}{2}(1 + tanh\frac{\mu}{2T}). \qquad (4.49)$$

Therefore

$$tanh\frac{\mu}{2T} \simeq n - 1 \qquad (4.50)$$

Substitution of Eq.(4.50,46) into Eq.(4.44) yields

$$\Sigma \simeq -(1 - n) \sum_{\mathbf{a}} \frac{T^2(\mathbf{a})}{2\omega_0 g^2(\mathbf{a})} \qquad (4.51)$$

or in the nearest neighbour approximation with the definition $D = zT(\mathbf{a})$

$$\Sigma \simeq -\frac{(1 - n)E_p}{2z\lambda^2} \qquad (4.52)$$

This expression is a result of the summation of the second order in H_{p-ph} multiphonon diagrams as shown in Fig.4.2a. The contribution of the second order being negative lowers the polaron energy and increases the effective mass further (to show this one should calculate the frequency derivative of $\Sigma(\omega)$). Gogolin (1982) in the framework of a single polaron problem estimated also the third and higher order contributions to Σ, Fig.4.2b

$$\Sigma^{(3)} \sim +\frac{E_p}{\lambda^3} \tag{4.53}$$

The third order contribution is positive and leads to the reduction of the effective mass. Because the dispersion is exponentially small one can sum all diagrams, including those with the intersection of the phonon lines. These results show that a consistent perturbation expansion in $1/\lambda$ exists with the small parameter

$$\frac{1}{2z\lambda^2} << 1 \tag{4.54}$$

where z is the nearest neighbour number. Therefore if the coupling constant $\lambda > 1/\sqrt{2z}$ small polarons are stable and they tunnel in a narrow band, Eq.(4.32).

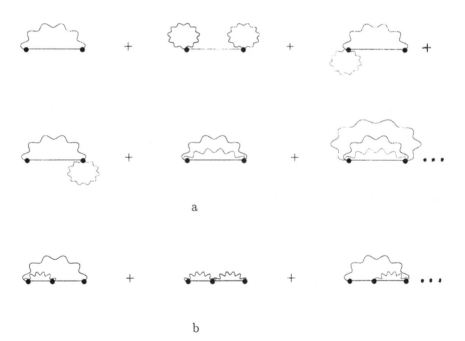

Fig.4.2. Polaron self-energy.

This condition of the small polaron formation is independent of the adiabatic ratio ω_0/D. However the term in front of the exponent in the bandwidth depends on this parameter. For the adiabatic small polaron $(\omega_0/D \ll 1)$ it is different from D as follows from the numerical calculations (see section 4.6).

4.5 Phonons in a strongly coupled system

An essential question which arises within the adiabatic Migdal approach is that of the phonon instability and the applicability of the Fröhlich hamiltonian. Taking into account the polaron formation we show in this section that the phonon frequency softening is small and therefore the Fröhlich hamiltonian is applicable also in the strong coupling regime.

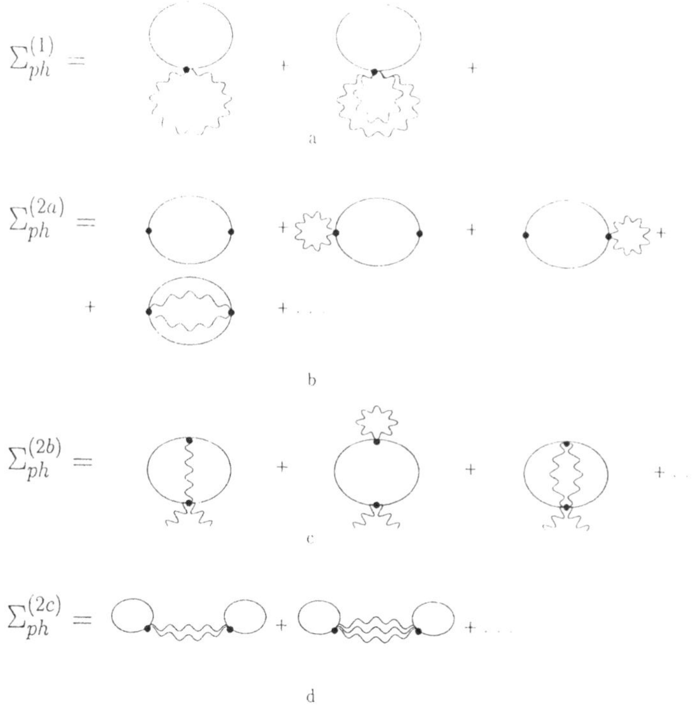

Fig. 4.3. Phonon self-energy in the strong coupling limit.

The first and different second order diagrams in H_{p-ph} contributing to the phonon self energy $\Sigma_{ph} = \omega_{\mathbf{q}}\Pi/2$ are shown in Fig.4.3a-d:

$$\Sigma_{ph} = \Sigma_{ph}^{(1)} + \Sigma_{ph}^{(2a)}(\mathbf{q}, \Omega_n) + \Sigma_{ph}^{(2b)}(\mathbf{q}) + \Sigma_{ph}^{(2c)}(\mathbf{q}). \qquad (4.55)$$

With exponential accuracy, $e^{-g^2} \ll 1$

$$\Sigma_{ph}^{(1)} = \Sigma_{ph}^{(2c)} = 0 \tag{4.56}$$

and

$$\Sigma_{ph} = -\frac{n(2-n)\gamma(\mathbf{q})^2}{2}\sum_{\mathbf{a}}(\Phi_{\mathbf{a}}(\Omega_n) - \Phi_{\mathbf{a}}(0))\left[1 - cos(\mathbf{q}\cdot\mathbf{a})\right]. \tag{4.57}$$

Calculating the Fourier component of the multiphonon correlator, Eq.(4.43) one obtains the main nonexponential contribution

$$\Sigma_{ph} = \frac{\Omega_n^2 n(2-n)\gamma(\mathbf{q})^2}{2\omega_0^3}\sum_{\mathbf{a}}\frac{T^2(\mathbf{a})[1 - cos(\mathbf{q}\cdot\mathbf{a})]}{g^6(\mathbf{a})} \tag{4.58}$$

The analytical continuation to the real frequencies is obtained by the simple substitution in Eq.(4.58)

$$i\Omega_n \rightarrow \omega + i0^+ \tag{4.59}$$

with the following result for the renormalised phonon frequency, which is a pole of the phonon GF:

$$\tilde{\omega}(\mathbf{q}) \simeq \omega_0 - \Delta(\mathbf{q}) \tag{4.60}$$

The phonon frequency softening

$$\Delta(\mathbf{q}) = \frac{n(2-n)\gamma^2(\mathbf{q})}{2\omega_0}\sum_{\mathbf{a}}\frac{T^2(\mathbf{a})[1 - cos(\mathbf{q}\cdot\mathbf{a})]}{g^6(\mathbf{a})} \tag{4.61}$$

is small compared with the frequency as $1/\lambda^2 \ll 1$ (Alexandrov and Capellmann (1991), Alexandrov *et al.* (1992)). Phonons are stable in the strong-coupling regime. The ions change their equilibrium positions due to the electron-phonon coupling retaining their vibration frequencies practically unchanged.

4.6 Transition from electron to small polaron

Several attempts to describe the intermediate region of the coupling $\lambda \simeq 1$ with and without electron-electron correlations are known in the literature. These are based on the variational approach (Eagles (1966), Emin (1973), Nasu (1985), Hang (1988),

Das and Sil (1989), Zheng *et al.* (1989,1990, 1994), Feinberg *et al.* (1990)), the Monte-Carlo calculations (De Raedt and Lagendijk (1983), Marsiglio (1990)), and on the exact (numerical) solution of a several-site model. The transition from a wide -band electron (or large polaron) to a narrow-band small polaron and a small bipolaron is extremely sharp as seen in the Monte-Carlo simulations, Fig.4.4. The general conclusion is that there is a continuous but rather sharp evolution from a wide-band electron to a narrow band small polaron (or bipolaron) in the intermediate region of the coupling $\lambda \simeq 1$.

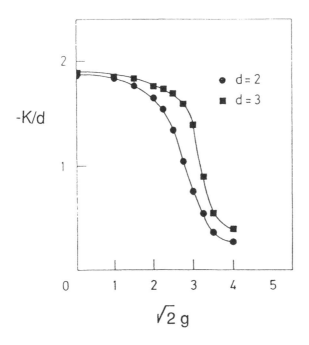

Fig.4.4. The 'Monte-Carlo' polaron collapse of the kinetic energy (K/t) of two-dimensional, $d = 2$, and three-dimensional, $d = 3$, fermions with increasing interaction constant g (de Raedt and Lagendijk (1983)).

The most reliable results for the intermediate region are obtained with the exact numerical diagonalization of vibrating clusters with one and two electrons (Kongeter and Wagner (1990), Ranninger and Thibblin (1992), Marseglio (1993), Alexandrov *et al.* (1994a)) . Numerical solution for several vibrating molecules coupled with one electron in the adiabatic $\omega/t < 1$ as well as in the nonadiabatic $\omega/t > 1$ regime using numerical diagonalization with 50 phonons shows that the adiabatic Holstein small polaron and the Lang-Firsov canonical transformation are in excellent agreement with the exact solution in adiabatic and nonadiabatic regimes correspondingly *for all values* of the coupling strength (Alexandrov *et al.* (1994a)). Here t is the value of the hopping integral between vibrating molecules. To illustrate this we compare in Fig.4.5 the exact splitting of levels with the expression (4.32) for nonadiabatic case and with the modified Holstein formula for two molecules in the adiabatic case:

$$w = \tilde{D}exp(-\tilde{g}^2) \qquad (4.62)$$

where

$$\tilde{D} = D\sqrt{\frac{8\lambda\omega}{\pi t}}\beta^{5/2}\lambda^{1-\beta}(2(1+\beta))^{-\beta}, \qquad (4.63)$$

with $\beta = \sqrt{1 - \frac{1}{4\lambda^2}}$, and

$$\tilde{g}^2 = g^2[\beta - \frac{1}{4\lambda}ln(2\lambda(1+\beta))] \qquad (4.64)$$

The expression Eq.(4.62) is a generalized Holstein formula, which works in the small polaron region, $\lambda > 1/2$ for the adiabatic case $\omega/t < 1$. We note that in the strong-coupling limit $\beta \simeq 1$. Therefore the exponent in the renormalized bandwidth, Eq.(4.62), is the same, as is obtained with the canonical transformation, Eq.(4.32): $\tilde{g} = g$. However, the term in front of the exponent, $\tilde{D}$, Eq.(4.63) differs from D for any value of λ. There is also an essential exponential difference between Eq.(4.32) and Eq.(4.62) in the intermediate coupling region, where $\tilde{g}$ differs from g. This leads to a much lower effective mass of the adiabatic small polaron in the intermediate coupling region compared with that estimated from the expression (4.32), Fig.4.5. In the intermediate coupling region, $\lambda \simeq 1$, where the Migdal approach is already meaningless, the effective bandwidth of a small adiabatic polaron ($\omega << t$) can be only one order of magnitude smaller than the bare band width, making the bipolaron high-T_c superconductivity quite feasible. That means also that one should sum a large number of diagrams in the $1/\lambda$ expansion (Section 4.4) to obtain a reliable expression for the effective mass of the adiabatic small polaron.

The renormalized phonon frequency has a minimum at the transition from large to small polaron as predicted with the Migdal formula, Fig.4.6. The solution of 4-site model shows also that at large λ the frequencies remain unchanged as follows from the $1/\lambda$ expansion.

4.7 Bipolaronic instability

Polarons are coupled not only with phonons via the residual interaction H_{p-ph} but also between themselves.

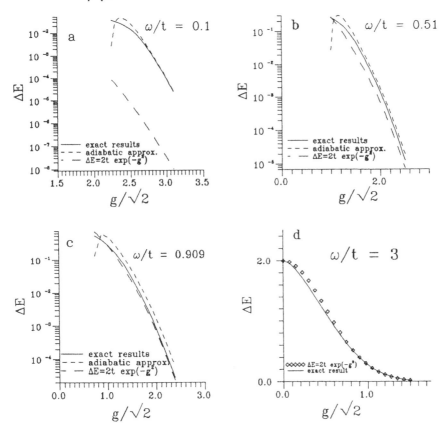

Fig.4.5. Band width of the two-molecular model (in units of t).

The effect of H_{p-p} is described with the dielectric response function $\epsilon(\mathbf{q}, \omega)$, for which the canonical random phase approximation can be adopted at sufficiently high temperature

$$\epsilon(\mathbf{q},\omega) = 1 - 2v(\mathbf{q}) \sum_{\mathbf{k}} \frac{n_{\mathbf{k+q}} - n_{\mathbf{k}}}{\omega - \epsilon_{\mathbf{k}} + \epsilon_{\mathbf{k+q}}}. \qquad (4.65)$$

One can apply this expression to describe the response of small polarons to a perturbation of a frequency $\omega < \omega_0$, when phonons in the polaronic cloud are not excited. Here $v(\mathbf{q})$ is the Fourier component of the polaron-polaron interaction

$$v(\mathbf{q}) = \frac{4\pi e^2}{\epsilon q^2} - \gamma^2(\mathbf{q})\omega_{\mathbf{q}} \qquad (4.66)$$

For the optical phonons in the long-wave limit

$$\gamma^2(\mathbf{q})\omega_0 = \frac{4\pi e^2 (\epsilon^{-1} - \epsilon_0^{-1})}{q^2} \qquad (4.67)$$

with ϵ_0 the static dielectric constant of a lattice, which takes into account the ion contribution to the polarisation.

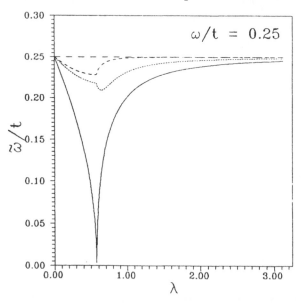

Fig.4.6. Phonon frequency renormalization as a function of the coupling constant for a 4-site cluster. All four phonon modes are shown. An asymmetric mode $(-u_1, -u_2, u_2, u_1)$ is unstable at $\lambda \simeq 0.57$.

Therefore at large distances the polaron-polaron interaction is the Coulomb repulsion:

$$v_{ij} = \frac{e^2}{\epsilon_0 |\mathbf{m} - \mathbf{n}|} \qquad (4.68)$$

in an ionic crystal and

$$v_{ij} = \frac{e^2}{\epsilon_c |\mathbf{m} - \mathbf{n}|} \qquad (4.69)$$

in an atomic or molecular solid. At short distances it might be repulsive or attractive depending on the value of a short range ϵ. In copper-based oxides the on-site Coulomb repulsion (Hubbard U) is normally large, of the order of several eV, and the attraction is possible only between carriers on the neighbouring (or next neighboring) sites. An estimate with the high frequency dielectric constant $\epsilon \simeq 4-5$ yields $V_c \simeq 0.5eV$ for the inter-site Coulomb repulsion. Therefore the inter-site attraction and the pairing of two polarons are possible if the short range electron-phonon interaction is strong enough: $E_p > V_c/2 \simeq 0.25eV$.

In the static limit at large distances (or $q \to 0$) we obtain the usual Debye screening due to the repulsion with the static dielectric function

$$\epsilon(q, 0) = 1 + \frac{q_s^2}{q^2} \qquad (4.70)$$

where $q_s = \sqrt{2\pi e^2 n(2-n)/T\epsilon_0}$. This result is obtained for a temperature large compared with the polaronic bandwidth using the expansion of the polaron distribution function:

$$n_{\mathbf{k}} \simeq \frac{n}{2}\left(1 - \frac{(2-n)\epsilon_{\mathbf{k}}}{2T}\right). \qquad (4.71)$$

The polaron response becomes dynamic already for a rather low frequency $\omega > w$:

$$\epsilon(\mathbf{q}, \omega) = 1 - \frac{\omega_p^2(\mathbf{q})}{\omega^2} \qquad (4.72)$$

with the temperature dependent plasma frequency

$$\omega_p^2(\mathbf{q}) = 2v(\mathbf{q})\sum_{\mathbf{k}} n_{\mathbf{k}}(\epsilon_{\mathbf{k}+\mathbf{q}} - \epsilon_{\mathbf{k}}), \qquad (4.73)$$

which is proportional to inverse temperature if $T >> w$. The expression (4.73) is applied to the polaron plasmon with a frequency below the optical phonon frequency, which is quite feasible due to a large value of the background dielectric constant and the enhanced effective mass. In the opposite case one should take into account the phonon shakeoff. If the phonon and polaron plasmon frequencies are close to each other a new type of vibration excitations exists due to the residual polaron-phonon coupling (Alexandrov (1992b)). They are a mixture of phonons and polaron plasmons, so called 'plasphons'.

For a short distance (large q) the Fourier component of the interaction $v(\mathbf{q}) = U$ might be positive or negative. The static dielectric function is given by ($T >> w$):

$$\epsilon(\mathbf{q}, 0) = 1 + \frac{n(2-n)U}{2T} \qquad (4.74)$$

A screened short-range interaction is given by

$$\tilde{U} = \frac{U}{\epsilon(\mathbf{q}, 0)} = \frac{UT}{T \pm T^{**}} \qquad (4.75)$$

with the characteristic temperature

$$T^{**} = \frac{|U|n(2-n)}{2}. \qquad (4.76)$$

The upper sign corresponds to the bare repulsion $U > 0$ while the lower sign ($-$) to the attraction, $U < 0$. One can see from Eq.(4.75) that in the temperature region $w < T << T^{**}$ the short range repulsion is sufficiently suppressed by the screening. The two-body correlations lead to a modification of polaron trajectories, which reduces the Coulomb self-energy to a magnitude of the order of the polaronic bandwidth if $T \simeq w$. In case of the attraction the singularity occurs in the effective interaction $\tilde{U}$ at $T = T^{**}$. This pole in the two-particle vertex part corresponds to the pairing of polarons. Therefore T^{**} in this case is the critical temperature of the *bipolaron* formation. For a half-filled band ($n \simeq 1$) T^{**} is of order of the attraction itself and might be as high as $10^3 K$.

4.8 Spin polarons and bipolarons

The small dielectric polarons treated above are formed by carriers in a conduction or valence band if the effective mass for a rigid band or the coupling with phonons are sufficiently large. If the undoped material is an antiferromagnetic insulator, a carrier can form another kind of polaron, a cluster of magnetic moments oriented ferromagnetically with orientation antiparallel to that of the carrier.

The relevance of spin polarons to high T_c materials arises from the fact that, in for instance the system $La_{2-x}Sr_xCuO_4$, the material for $x = 0$ is an antiferromagnetic insulator, and for finite value of x can be treated as doped with strontium acting as accepter centres. The spin polaron being heavy certainly polarises the lattice in polar material, so a hybrid pseudoparticle is likely to form, both spin and lattice polarisation contribute to the mass enhancement, but only the second one to the isotope effect.

We next reproduce an estimate of the conditions under which a spin polaron is formed. The interaction between an electron spin **s** on a carrier in a conduction or valence band (c) and the moment on a d or f like orbital responsible for the magnetism can be written as

$$J_{cf}\mathbf{s} \cdot \mathbf{S} \tag{4.77}$$

where J_{cf} is the Hund's rule energy coupling the spin **s** of the conduction electron and the moment **S** on the ion responsible for antiferromagnetism. The spin polaron consists of a sphere of radius r_p round a carrier in which the moments are polarised ferromagnetically in the direction antiparallel to **s**. The conduction electron is then shut up in a spherical box of this radius, which costs the kinetic energy $\pi^2/2mr_p^2$. Let J_N be the energy per moment needed to go from the antiferromagnetic to the ferromagnetic state due to interaction with neighbors. Then the energy of the polaron is

$$\frac{\pi^2}{2mr_p^2} + \frac{4\pi}{3}\left(\frac{r_p}{a}\right)^3 J_N - J_{cf} \tag{4.78}$$

This is a minimum when

$$r_p^5 = \frac{\pi a^3}{4mJ_N} \tag{4.79}$$

Introducing an effective coupling constant $\lambda = J_N/(1/2ma^2)$ one can rewrite this expression as $r_p \simeq a/\lambda^{1/5}$. The total energy is

$$\frac{5\pi^2}{6m}\left(\frac{4mJ_N}{\pi a^3}\right)^{2/5} - J_{cf} \qquad (4.80)$$

A spin polaron forms if this is negative. Fig.4.7a shows how we envisage a spin polaron moving in an antiferromagnetic material above the Néel temperature.

Two spin polarons form a singlet spin bipolaron, Fig 4.7b, if the exchange energy J_N is sufficiently large (Mott (1990)).

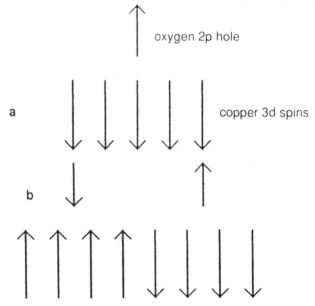

Fig.4.7. Spin polaron (a) and spin bipolaron (b).

The binding energy to form a singlet bipolaron will be of order

$$J_N\left(\frac{r_p}{a}\right)^{d-1} \qquad (4.81)$$

where d is the dimensionality ($d = 1, 2$ or 3).

Below the Néel temperature (T_N) we suppose that a spin polaron moves in a band with a k-value like a small polaron below the Debye temperature, with rather strong scattering by magnons. To obtain the effective mass m^* we compute the transfer integral when the polaron moves through one atomic distance. The spins will contribute to the transfer integral a term proportional

$$\prod_N cos\theta_N \qquad (4.82)$$

where θ_N is the change in orientation when the carrier moves through one atomic distance. We may expect that

$$\theta_N \sim \frac{a}{r_p}. \qquad (4.83)$$

So Eq.(4.82) becomes

$$\prod_{N=1}^{N_{max}} \left(1 - \frac{a^2}{2r_p^2}\right)^N \qquad (4.84)$$

with $N_{max} \simeq (r_p/a)^3$. For large r_p this varies as

$$exp\left(-const\frac{r_p}{a}\right) \qquad (4.85)$$

and proportional to $1/m^*$. Therefore the spin polaron mass increases exponentially with the size of the spin polaron, which is opposite to the case of the lattice polaron.

For $T > T_N$, we expect the motion to be diffusive. Vigren (1973) suggested that the diffusion coefficient should be independent of temperature. The argument is that the polaron will move as a consequence of a spin flip at its perimeter, so that the diffusion coefficient would be

$$D \sim \frac{a^2}{T_N} \qquad (4.86)$$

because the frequency of a spin flip is of the order T_N. If the same is valid for spin bipolarons, which are nondegenerate above T_c, then the electrical conductivity is inversely proportional to temperature

$$\sigma = \frac{4e^2 n D}{T} \qquad (4.87)$$

and there is no residual resistivity (Mott (1991)).

The strongest evidence for spin polarons at the present time is provided by the work of von Molnar *et al.* (1983) on $Cd_{3-x}v_xS_4$, where v stands for vacancy, x of order 0.1 in these materials, this being the concentration of electrons in the conduction band. Magnetic properties suggest that the polaron contains about 20 moments. The most striking evidences comes from the negative magneto-resistance. The explanation given is that the polaron is so heavy that at low temperature it suffers Anderson localisation through interaction with the random field of the vacancies. A field of 20 kOe begins to destroy the polarons, producing the ferromagnetic saturation of the moments: the conductivity begins to rise.

One can compare the properties of lattice and spin polarons in more detail performing the Holstein-Primakoff transformation for the spin part of the total hamiltonian (see for review Lu Yu *et al.* (1993)). It turns out that the exchange integral J_N is, to some extent, equivalent to the phonon energy ω in the lattice polaron problem, and the spin distortion caused by the hole is equivalent to the lattice one. However magnons are not perfect bosons, and this fact makes some properties of spin polarons (e.g. effective mass) quite different from those of lattice polarons. The further complication appears in copper based oxides due to the substantial hybridisation of the oxygen p and copper d orbitals. As a result an oxygen hole hops through the copper sites at which spins are localised. In the large Hubbard U limit the appropriate hamiltonian is the so-called $t-J$ model proposed by several authors. As argued by Zhang and Rice (1988) the low-energy physics of CuO_2 plane of copper based oxides is determined by the singlet state formed by the oxygen hole with the copper spin. Nevertheless even in this case the idea that only the linearised collective excitations (magnons) of the spin system matter turns out to be constructive and the problem of the carrier strongly coupled with the spin background can be treated with the powerful polaron formalism (Schmitt-Rink *et al.* (1988)).

We believe that the cooperative properties of spin and lattice bipolarons are those of charged bosons and can be described within the same formalism (Chapter 5). More details about large and small bipolarons and their cooperative properties will be found in our review 'Bipolarons' (Alexandrov A S and Mott N F 1994 Reports on Progress in Physics (IOP), to be published).

Chapter 5

Strong coupling superconductivity

The transition from wide -band electrons (or large polarons) to small polarons is extremely sharp as discussed in Chapter 4. Nevertheless the evolution from the Fermi liquid to the charged Bose liquid might vary smoothly with the *polaronic* BCS -like superconductivity in a very narrow region of the coupling constant. The latter has high T_c due to the large density of states in the extremely narrow (less than $10^3 K$) polaronic band. Therefore we first consider the polaronic superconductivity, which is realised if the Coulomb repulsion almost compensates the attraction.

5.1 Polaronic superconductivity

5.1.1 Cooper pairing of nonadiabatic carriers

The polaron-polaron interaction is the sum of two large contributions of the opposite sign (Eq.(4.22)) which generally are large compared with the polaron bandwidth. This is just the opposite regime to that of the BCS superconductor where the Fermi energy is the largest one. However, for an intermediate value of λ the effective interaction of polarons can be comparable or even less than the polaron bandwidth, both being less or of order of the characteristic phonon frequency. This is a narrow region of the coupling where the BCS approach is applied with a *nonretarded* attraction and Cooper pairs formed by two *small polarons* (Alexandrov

(1983)). The appropriate hamiltonian is the extended Hubbard hamiltonian taking into account the polaron narrowing of the band

$$H_p = \sum_{i,j} \left(\sigma(\mathbf{m} - \mathbf{n}) - \mu \delta_{i,j} \right) c_i^\dagger c_j + \frac{1}{2} v_{ij} c_i^\dagger c_j^\dagger c_j c_i \right) \qquad (5.1)$$

where $\sigma(\mathbf{m})$ is determined with Eq.(4.32) for a nonadiabatic system $(\omega > D/z)$, or with Eq.(4.62) in the opposite case. For simplicity one can keep only the on-site v_0 and the nearest neighbor intersite v_1 interactions. At least one of them should be attractive to ensure the superconducting ground state.

If a pair binding energy Δ is small compared with the renormalised bandwidth $2w$, polarons constituting a pair tunnel through many sites during the characteristic time $1/\Delta$. Therefore a pair spreads over a large number of sites and bipolarons are overlapped similar to Cooper pairs. By introducing two order parameters

$$\Delta_0 = -v_0 \langle c_{\mathbf{m},\uparrow} c_{\mathbf{m},\downarrow} \rangle \qquad (5.2)$$

$$\Delta_1 = -v_1 \langle c_{\mathbf{m},\uparrow} c_{\mathbf{m}+\mathbf{a},\downarrow} \rangle \qquad (5.3)$$

and transforming to the $\mathbf{k}$ -representation one arrives at the usual BCS hamiltonian

$$H_p = \sum_{\mathbf{k},s} \left(\xi_\mathbf{k} c_{\mathbf{k},s}^\dagger c_{\mathbf{k},s} + [\Delta(\mathbf{k}) c_{\mathbf{k},\uparrow}^\dagger c_{-\mathbf{k},\downarrow}^\dagger + h.c.] \right) \qquad (5.4)$$

where $\xi_\mathbf{k} = \epsilon_\mathbf{k} - \mu$ is the kinetic energy of the polaron tunneling with momentum $\mathbf{k}$ and

$$\Delta(\mathbf{k}) = \Delta_0 - \Delta_1 \frac{\xi_\mathbf{k} + \mu}{w} \qquad (5.5)$$

is the order parameter corresponding to a singlet pairing. In general a triplet p-wave pairing are possible with the hamiltonian Eq.(5.1).

Applying the standard diagonalization procedure to the hamiltonian Eq.(5.1) one obtains for the order parameter

$$\langle c_{\mathbf{k},\uparrow} c_{-\mathbf{k},\downarrow} \rangle = \frac{\Delta(\mathbf{k})}{2\sqrt{\xi_\mathbf{k}^2 + \Delta(\mathbf{k})^2}} tanh \frac{\sqrt{\xi_\mathbf{k}^2 + \Delta(\mathbf{k})^2}}{2T} \qquad (5.6)$$

and with the definition Eq.(5.2,3)

$$\Delta_0 = -\frac{v_0}{N}\sum_{\mathbf{k}}\frac{\Delta(\mathbf{k})}{2\sqrt{\xi_{\mathbf{k}}^2 + \Delta(\mathbf{k})^2}}tanh\frac{\sqrt{\xi_{\mathbf{k}}^2 + \Delta(\mathbf{k})^2}}{2T} \tag{5.7}$$

$$\Delta_1 = -\frac{v_1}{Nw}\sum_{\mathbf{k}}\frac{\Delta(\mathbf{k})(\xi_{\mathbf{k}} + \mu)}{2\sqrt{\xi_{\mathbf{k}}^2 + \Delta(\mathbf{k})^2}}tanh\frac{\sqrt{\xi_{\mathbf{k}}^2 + \Delta(\mathbf{k})^2}}{2T} \tag{5.8}$$

The last two equations are equivalent to the BCS one for $\Delta(\mathbf{k}) = \Delta(\xi)$ with the potential depending on energy and with the half band width w as a cutoff of the integral instead of the Debye temperature:

$$\Delta(\xi) = \int_{-w-\mu}^{w-\mu} d\xi' N_p(\xi')V(\xi,\xi')\frac{\Delta(\xi')}{2\sqrt{\xi'^2 + \Delta(\xi')^2}}tanh\frac{\sqrt{\xi'^2 + \Delta(\xi')^2}}{2T} \tag{5.9}$$

with $V(\xi,\xi') = -v_0 - zv_1\frac{(\xi+\mu)(\xi'+\mu)}{w^2}$. The polaron density of states is enhanced due to the polaron narrowing effect:

$$N_p(\xi) \equiv \frac{1}{N}\sum_{\mathbf{k}}\delta(\xi - \xi_{\mathbf{k}}) \tag{5.10}$$

being of order of $1/2w$ instead of $1/2D$ for a bare band. The on-site and intersite interactions are both attractive (< 0) or one of them (on-site) may be repulsive. The critical temperature T_c is determined from the two linearised equations (5.7,8) in the limit $\Delta_{0,1} \to 0$:

$$\left(1 + A(\frac{v_0}{zv_1} + \frac{\mu^2}{w^2})\right)\Delta - \frac{B\mu}{w}\Delta_1 = 0 \tag{5.11}$$

$$-\frac{A\mu}{w}\Delta + (1 + B)\Delta_1 = 0 \tag{5.12}$$

where $\Delta = \Delta_0 - \Delta_1\frac{\mu}{w}$ and

$$A = \frac{zv_1}{2w}\int_{-w-\mu}^{w-\mu}\frac{d\xi tanh\frac{\xi}{2T_c}}{\xi} \tag{5.13}$$

$$B = \frac{zv_1}{2w}\int_{-w-\mu}^{w-\mu}\frac{d\xi\xi tanh\frac{\xi}{2T_c}}{w^2}. \tag{5.14}$$

These equations are applied only for a weak and intermediate polaron-polaron coupling $v_{0,1} < w$. In the limit of the weak coupling one obtains from Eq.(5.11,12)

$$T_c \simeq 1.14w\sqrt{1 - \frac{\mu^2}{w^2}}exp\left(\frac{2w}{v_0 + zv_1\frac{\mu^2}{w^2}}\right) \tag{5.15}$$

The expression Eq.(5.15) plays the same role in the polaronic superconductivity as the BCS one in the low temperature superconductors. It predicts superconductivity even in the case of on-site repulsion $v_0 > 0$ if this repulsion is less than the total intersite attraction $z|v_1|$. It also predicts a nontrivial dependence of T_c on the doping. With the constant density of states within the polaron band the Fermi level μ is expressed through the number of polarons per atom n

$$\mu = w(n - 1) \tag{5.16}$$

and

$$T_c \simeq 1.14w\sqrt{n(2 - n)}exp\left(\frac{2w}{v_0 + zv_1(n - 1)^2}\right) \tag{5.17}$$

T_c has two maxima as a function of n separated by a deep minimum for the half filled band ($n = 1$) when the nearest neighbor contributions to the pairing are mutually compensated.

5.1.2 High T_c

The basic phenomenon that allows the high value of T_c is that the polaronic narrowing of the band, which eliminates the small exponential factor in the BCS or McMillan's formula Eq(3.73). To show this we rewrite Eq.(5.17) in a slightly different form separating the phonon mediated attraction and the Coulomb repulsion and taking explicitly into account the polaronic narrowing effect:

$$T_c \simeq \tilde{D}exp\left(-g^2 - \frac{exp(-g^2)}{\lambda - \mu_c}\right) \tag{5.18}$$

where $\tilde{D} = 1.14D\sqrt{n(2 - n)}$,

$$\lambda = \frac{2E_p + z(n - 1)^2 \sum_{m=a} \gamma^2(q)\omega_q e^{iq \cdot m}}{2D}, \tag{5.19}$$

and

$$\mu_c = \frac{U + z(n-1)^2 V_c}{2D} \qquad (5.20)$$

with U and V_c the onsite and intersite Coulomb repulsion respectively.

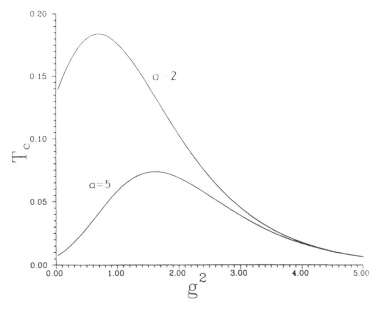

Fig.5.1. Critical temperature of a polaronic superconductor (in units of $\tilde{D}$) as a function of the interaction constant g^2 for two different values of the attraction between polarons, $a = 1/(\lambda - \mu_c)$.

There are four independent parameters which determines the value of T_c, in particular the bare bandwidth D, the polaronic level shift E_p, the number of phonons g^2 in a polaronic cloud, and the Coulomb repulsion μ_c. They correspond to the four independent parameters of the Fröhlich hamiltonian: the electron kinetic energy $E_F \sim D$, the matrix element of the electron-phonon interaction $\gamma \sim g$, the characteristic phonon frequency $\omega = E_p/g^2$, and the Coulomb (pseudo)potential. Because only g^2 (or ω) depends on the ion mass one can determine the maximum value of T_c with respect to g^2 keeping D, λ and μ constant. Differentiating Eq.(5.18) with respect to g^2 we obtain for the maximum T_c^*

$$T_c^* = \tilde{D}\frac{\lambda - \mu_c}{e} \qquad (5.21)$$

The applicability of this formula is restricted by the intermediate region of the interaction constant $\lambda - \mu_c < 1$ because of the formation of bipolarons in the large λ limit. However if bipolarons are intersite and their effective mass is of the order of the polaron effective mass the expression Eq.(5.18) also describes the Bose-Einstein condensation of bipolarons, as we discuss in Chapter 7. T_c^* is limited by the condition of the small polaron formation $\lambda > 1/\sqrt{2z}$ (Section 4.4), which restricts the maximum value of D in Eq.(5.21). The value of g^2 at which T_c reaches its maximum is $g^2 = \ln(\lambda - \mu_c)^{-1}$ so

$$D < \sqrt{2z}\omega \ln(\lambda - \mu_c)^{-1} \qquad (5.22)$$

and

$$T_c^* < \omega \frac{\sqrt{2z}(\lambda - \mu_c)\ln(\lambda - \mu_c)^{-1}}{e} \qquad (5.23)$$

The maximum value of the critical temperature of order of $\omega/3$ is reached in the region of the intermediate coupling $g^2 \simeq 1$, Fig.5.1, where on the contrary the BCS superconductor has rather low T_c of order of 0.1ω. The absolute value depends on ω and can be as high as $200 - 300K$ depending on the value of the high frequency optical mode. At large value of the coupling T_c for a bipolaronic superconductor drops because carriers become very heavy. Therefore we conclude that the highest T_c is in the transition region from polaronic to bipolaronic superconductivity. The fact that due to the polaron nonadiabaticity the short-range Coulomb pseudopotential μ_c is not suppressed contrary to the BCS case with μ_c^* rather than μ_c does not change this conclusion if $\lambda - \mu_c$ is positive.

5.1.3   Normal state of a polaronic superconductor

In the normal state for sufficiently high temperature $(T > T^{**})$ the polaron-polaron correlations are not very important and in the first approximation the polaron heat capacity as well as the magnetic susceptibility are those of narrow band fermions. Moreover in the temperature range $T < \omega$ the polaron bandwidth is temperature independent. We assume here that the characteristic phonon frequency is high: $\omega > T^{**}$. For a narrow band Fermi-gas one obtains the heat capacity C_p:

$$C_p = 2T \int_{-w-\mu}^{w-\mu} d\xi N_p(\xi) \frac{df}{d\xi} \left(\frac{\xi}{T} \frac{d\mu}{dT} - \frac{\xi^2}{T^2} \right), \qquad (5.24)$$

where $f(\xi) = \left(exp(\frac{\xi}{T}) + 1 \right)^{-1}$. This expression provides a linear temperature dependence of C_p for low temperatures $T < 0.4w$ if the polaron density of states is energy independent $N_p = 1/2w$:

$$C_p = \frac{\pi^2 T}{3w} \qquad (5.25)$$

and a power law decrease $(\sim T^{-2})$ for $T > w$:

$$C_p = \frac{w^2 n(2 - n)}{6T^2} \qquad (5.26)$$

The numerical calculations in the intermediate temperature region reveals only a small change in the position T_m of the maximum of $C_p(T)$ with the variation of the filling factor n, $T_m \simeq 0.4w$, and a gradual increase of the maximum value from $C_p^m = 0.2$ at $n = 0.2$ to $C_p^m = 0.6$ at $n = 1$. The temperature dependence of the heat capacity is similar to the Schottky anomaly except for the low temperature region with the linear C_p instead of the exponential one of two-level systems.

In the polaronic system the narrow band includes all states of the Brillouin zone rather than a small part of them under a peak of the density of states as in a so-called 'van -Hove scenario' of high-T_c. This fact provides us with the possibility to get an absolute value of the spin susceptibility of polarons considerably higher than those of wide band electrons

$$\chi_s = -\frac{\mu_B^2}{w} \int_{-w-\mu}^{w-\mu} d\xi \frac{df}{d\xi} \qquad (5.27)$$

where μ_B is the Bohr magneton. The chemical potential is determined through the density of polarons

$$\mu = T \ln \frac{exp\frac{nw}{T} - 1}{exp\frac{w}{T} - exp\frac{(n-1)w}{T}} \qquad (5.28)$$

and

$$\chi_s = \frac{\mu_B^2}{w} \frac{(exp\frac{nw}{T} - 1)(exp\frac{(2-n)w}{T} - 1)}{exp\frac{w}{T} - 1} \qquad (5.29)$$

Eq.(5.29) yields the Curie law in the high-temperature limit, $T \gg w$

$$\chi_s \simeq \frac{\mu_B^2 n(2-n)}{2T} \qquad (5.30)$$

and a temperature independent susceptibility enhanced due to the polaron narrowing of the band for $T < w$:

$$\chi_s = \frac{\mu_B^2}{w} \qquad (5.31)$$

with w instead of D as in ordinary metals. This enhancement is the same as that of the specific heat ratio $\gamma = C_p/T$ Eq.(5.25), which is quite different from that in ordinary metals, where only the electronic heat capacity is increased by the electron-phonon interaction while the magnetic susceptibility remains unaffected .

Transport properties of polarons depend strongly on the temperature range. For temperature lower than the Debye temperature (or the characteristic phonon energy) polarons tunnel through the narrow band, however at higher temperature the polaronic band collapses and transport is diffusive via thermally activated hopping (see e.g. Mott and Davis (1979)).

To conclude the discussion of the polaron systems we notice that many unusual properties of 'old' high-T_c materials ($T_c < 20K$) such as $A-15$ and $C-15$ compounds were understood with the polaron band narrowing (for review see Alexandrov and Elesin (1985)). The polaronic superconductivity can be realised in highly ionic materials with a large static dielectric constant ϵ_0. In those oxides the long-range Coulomb repulsion is almost compensated by the lattice polarization, so the short-range residual attraction can be rather small while the electron-phonon coupling is sufficient to form small polarons. Some high-T_c oxides like $BSSCO$ in which a well defined Fermi-surface seems to be measured may be polaronic while others like $YBCO$ are bipolaronic (for more details concerning this difference see Chapter 7).

5.2 Bipolaronic superconductivity

5.2.1 Bipolaron band and repulsion

As we have mentioned the on-site or intersite attractive energy of two small polarons is generally larger than the polaron bandwidth. At this condition real space pairs of polarons (small bipolarons) form. Their properties are those of charged bosons. A small parameter $w/\Delta \ll 1$ where Δ is the bipolaron binding energy, of order of the attractive energy provides us with a consistent treatment of the bipolaronic systems (Alexandrov and Ranninger (1981a,b)). Under this condition the hopping term in the transformed hamiltonian $\tilde{H}$ is a small perturbation to the ground state of immobile bipolarons and free phonons:

$$\tilde{H} = H_0 + H_{pert} \tag{5.32}$$

where

$$H_0 = \frac{1}{2} \sum_{i,j} v_{ij} c_i^\dagger c_j^\dagger c_j c_i + \sum_{\mathbf{q}} \omega_{\mathbf{q}} d_{\mathbf{q}}^\dagger d_{\mathbf{q}} \tag{5.33}$$

and

$$H_{pert} = \sum_{i,j} \hat{\sigma}_{ij} c_i^\dagger c_j \tag{5.34}$$

Under the condition $\Delta \gg w$ there are no unbound polarons in the ground state, and the bipolaron motion can be described with a new canonical transformation $exp(S_2)$. This transformation eliminates the first order of H_{pert}, which destroys a bipolaron and therefore has no diagonal contribution:

$$(S_2)_{f,p} = \sum_{i,j} \frac{\langle f | \hat{\sigma}_{ij} c_i^\dagger c_j | p \rangle}{E_f - E_p}, \tag{5.35}$$

where $E_{f,p}$ and $|f\rangle, |p\rangle$ are the energy levels and the eigenstates of H_0. Neglecting the terms higher order than $(w/\Delta)^2$ one obtains

$$(H_b)_{ff'} = \left(e^{S_2} \tilde{H} e^{-S_2} \right)_{ff'} \tag{5.36}$$

$$(H_b)_{ff'} \simeq (H_0)_{ff'} - \frac{1}{2} \sum_{\nu} \sum_{i,i';j,j'} \langle f | \hat{\sigma}_{ii'} c_i^\dagger c_{i'} | \nu \rangle \langle \nu | \hat{\sigma}_{jj'} c_j^\dagger c_{j'} | f' \rangle$$

$$\times \quad \frac{E_f + E_{f'} - 2E_\nu}{(E_f - E_\nu)(E_\nu - E_{f'})}. \tag{5.37}$$

The nonzero matrix elements of S_2 act between a localised bipolaron state and a state of two independent polarons localised in different cells or sites (in case of on-site pairs). The expression (5.37) determines the matrix elements of the transformed (bipolaronic) hamiltonian H_b in the subspace $|f\rangle, |f'\rangle$ without real polarons. On the other side $|\nu\rangle$ refers to configurations involving two unpaired polarons, so that

$$E_f - E_\nu = -\Delta + \sum_\mathbf{q} \omega_\mathbf{q} \left(n_\mathbf{q}^f - n_\mathbf{q}^\nu \right), \tag{5.38}$$

where $n_\mathbf{q}^{f,\nu}$ are the phonon occupation numbers $(0,1,2,3...)$. This equation is an explicit definition of the bipolaron binding energy Δ, which takes into account the interaction between bipolarons as well as between two unpaired polarons. The lowest eigenstates of H_b are in the subspace which involves only doubly occupied $c_{m,s}^\dagger c_{m,s'}^\dagger |0\rangle$ or empty $|0\rangle$ states. The bipolaron tunneling takes place via a transition to a virtual unpaired state implying a single polaron tunneling to the adjacent cell. The subsequent tunneling of a second polaron of a pair restores the initial energy state of the system. Because the bipolaron band is narrow (see below) there are no high-frequency *real* phonons emitted or absorbed. Hence one can average H_b with the phonon density matrix

$$H_b = H_0 - i \sum_{i,i';j,j'} c_i^\dagger c_{i'} c_j^\dagger c_{j'} \int_0^\infty dt e^{-i\Delta t} \Phi_{ii'}^{jj'}(t) \tag{5.39}$$

with the real time t in the multiphonon correlator Φ determined with Eq.(4.42). Taking into account that there are only bipolarons in the subspace in which H_b operates one can rewrite the bipolaron hamiltonian in terms of the creation $b_i^\dagger = c_{m\uparrow}^\dagger c_{m,\downarrow}^\dagger$ and annihilation singlet bipolaron operators

$$\begin{aligned} H_b &= -\sum_\mathbf{m} \left(\Delta + \frac{1}{2} \sum_{\mathbf{m}'} v_{\mathbf{m},\mathbf{m}'}^{(2)} \right) n_\mathbf{m} \\ &+ \sum_{\mathbf{m} \neq \mathbf{m}'} \left(-t_{\mathbf{m},\mathbf{m}'} b_\mathbf{m}^\dagger b_{\mathbf{m}'} + \frac{1}{2} \bar{v}_{\mathbf{m},\mathbf{m}'} n_\mathbf{m} n_{\mathbf{m}'} \right), \tag{5.40} \end{aligned}$$

where $n_{\mathbf{m}} = b_{\mathbf{m}}^{\dagger} b_{\mathbf{m}}$ is the bipolaron occupation number operator,

$$\bar{v}_{\mathbf{m},\mathbf{m}'} = 4v_{\mathbf{m},\mathbf{m}'} + v^{(2)}_{\mathbf{m},\mathbf{m}'}, \qquad (5.41)$$

which is the bipolaron-bipolaron interaction including the direct (density-density) Coulomb repulsion (V_c), the attraction via phonons between two small polarons in different cells and a second order correction

$$v^{(2)}_{\mathbf{m},\mathbf{m}'} = 2i \int_0^{\infty} dt \Phi^{\mathbf{m'm}}_{\mathbf{mm'}}(t) exp(-i\Delta t), \qquad (5.42)$$

which is repulsive. The origin of this repulsion is quite clear: a virtual hop of one of the polarons of a pair is forbidden when the neighboring cell is occupied by another pair. The bipolaron transfer integral is of the second order in the electron kinetic energy $T(\mathbf{m})$ (see Fig. 4.1):

$$t_{\mathbf{m},\mathbf{m}'} = 2i \int_0^{\infty} dt \Phi^{\mathbf{mm'}}_{\mathbf{mm'}}(t) exp(-i\Delta t). \qquad (5.43)$$

To calculate t and $v^{(2)}$ one should use the explicit form of the multiphonon correlator. For the real time

$$\Phi^{\mathbf{mm'}}_{\mathbf{mm'}}(t) = T^2(\mathbf{a}) e^{-2g^2} exp \left(\frac{1}{N} \sum_{\mathbf{q}} \gamma^2(\mathbf{q}) f_{\mathbf{q}} \frac{cos\left(\omega_{\mathbf{q}}[\frac{i}{2T}+t]\right)}{sinh \frac{\omega_{\mathbf{q}}}{2T}} \right) \qquad (5.44)$$

with $f_{\mathbf{q}} = cos(\mathbf{q} \cdot \mathbf{a}) - 1$ and the opposite sign of the argument of the second exponent in the case of $\Phi^{\mathbf{m'm}}_{\mathbf{mm'}}$. For $T = 0$ and dispersionless phonons this yields

$$t_{\mathbf{mm'}} = \frac{2T^2(\mathbf{a})}{\Delta} e^{-2g^2} \sum_{k=0}^{\infty} \frac{(-2g^2)^k}{k!(1+k\omega/\Delta)} \qquad (5.45)$$

and

$$v^{(2)}_{\mathbf{mm'}} = \frac{2T^2(\mathbf{a})}{\Delta} e^{-2g^2} \sum_{k=0}^{\infty} \frac{(+2g^2)^k}{k!(1+k\omega/\Delta)} \qquad (5.46)$$

with $\mathbf{a} = \mathbf{m} - \mathbf{m}'$. If $\Delta < \omega$ both the bipolaron hopping and the second order repulsion are about w^2/Δ. However for large binding energy $\Delta \gg \omega$ the bipolaron bandwidth dramatically decreases being proportional to e^{-4g^2} in the limit $\Delta \to \infty$, and on the contrary their repulsion increases, $v^{(2)} \sim D^2/\Delta$ in this limit. For dispersionless molecular phonons the intermolecular attraction via phonons (the second term in Eq.(4.22)) is zero. This together with the direct Coulomb and second order $v^{(2)}$ repulsive terms lead to a total repulsive interaction between bipolarons, which prevents the formation of droplets.

In case of inter-site bipolarons there are two additional points. First of all an inter-site bipolaron can be formed with a nonzero spin $S = 1$ (triplet state) and with the energy $J \sim w^2/U$ above the singlet state $S = 0$. This should be taken into account by introducing the additional spin quantum numbers $S = 1; l = 0, \pm 1$ in the definition of $b_{\mathbf{m}}$. The second point is that in a simple square or cubic lattice the inter-site bipolaron tunnels via the next neighbor hopping of a single polaron rather than via two-particle (Josephson-like) tunneling, described with Eq.(5.43). This 'crab-like' tunneling results in a bipolaron bandwidth of the same order as the polaron one. Therefore H_b in the form (5.40) is applied not only to on-site bipolarons but also to intersite or more extended nonoverlapping pairs if $\mathbf{m}$ includes the spin, and $t_{\mathbf{m},\mathbf{m}'}$ is considered as a phenomenological parameter. The site index $\mathbf{m}$ should be generally considered as a position of the centre of mass of the bipolaron.

5.2.2 Ground state and excitations of a bipolaronic liquid

We rewrite the bipolaronic hamiltonian for a perfect lattice in the form

$$H_b = -\mu \sum_{\mathbf{m}} n_{\mathbf{m}} + \sum_{\mathbf{m} \neq \mathbf{m}'} \left(\frac{1}{2} \bar{v}_{\mathbf{m}\mathbf{m}'} n_{\mathbf{m}} n_{\mathbf{m}'} - t_{\mathbf{m}\mathbf{m}'} b_{\mathbf{m}}^\dagger b_{\mathbf{m}'} \right) \quad (5.47)$$

where the energy of a single *localised* bipolaron is included in the definition of the *bipolaron* chemical potential μ. The bipolaron operators obey the mixed commutation rules in a subspace of empty or doubly occupied sites (cells)

$$b_{\mathbf{m}} b_{\mathbf{m}}^\dagger + b_{\mathbf{m}}^\dagger b_{\mathbf{m}} = 1, \quad (5.48)$$

and

$$b_{\mathbf{m}} b_{\mathbf{m}'}^\dagger - b_{\mathbf{m}'}^\dagger b_{\mathbf{m}} = 0 \quad (5.49)$$

for $\mathbf{m} \neq \mathbf{m}'$. This makes useful the pseudospin analogy

$$b_{\mathbf{m}}^{\dagger} = S_{\mathbf{m}}^{x} - iS_{\mathbf{m}}^{y} \qquad (5.50)$$

and

$$b_{\mathbf{m}}^{\dagger} b_{\mathbf{m}} = \frac{1}{2} - S_{\mathbf{m}}^{z} \qquad (5.51)$$

with the spin(pseudo) 1/2 operators $S^{x,y,z} = \frac{1}{2}\tau_{1,2,3}$. $S^{z} = 1/2$ corresponds to an empty cell and $S^{z} = -1/2$ to a cell occupied by a bipolaron. The spin operators preserve the bosonic character of bipolarons if they are on different cells and their fermionic (or hard core) internal structure. Replacing bipolarons for the spin operators we transform the bipolaronic hamiltonian Eq.(5.47) into the familiar anisotropic Heisenberg hamiltonian

$$H_b = \mu \sum_{\mathbf{m}} S_{\mathbf{m}}^{z} + \sum_{\mathbf{m} \neq \mathbf{m}'} \left(\frac{1}{2}\bar{v}_{\mathbf{mm}'} S_{\mathbf{m}}^{z} S_{\mathbf{m}'}^{z} - t_{\mathbf{mm}'}(S_{\mathbf{m}}^{x} S_{\mathbf{m}'}^{x} - S_{\mathbf{m}}^{y} S_{\mathbf{m}'}^{y}) \right)$$

$$(5.52)$$

with the bipolaron chemical potential playing the role of an external magnetic field. This hamiltonian has been investigated in detail as a relevant form for magnetism and also for quantum solids like a lattice model for He^4. However , while in those cases the magnetic field is an independent thermodynamic variable, in our case it is fixed by the total 'magnetization' because the bipolaron density n is conserved

$$\frac{1}{N} \sum_{\mathbf{m}} \langle\langle S_{\mathbf{m}}^{z} \rangle\rangle = \frac{1}{2} - n \qquad (5.53)$$

That leads to an important difference in the phase diagram and the excitation spectrum. For the ground state one can apply a mean field approach introducing an average magnetic field $\mathbf{H_m}$ acting on a spin $\mathbf{m}$. In the nearest neighbor approximation

$$\mathbf{H_m} = -(\mu + 2\bar{v}\langle S_{\mathbf{m}'}^{z} \rangle)\mathbf{e} + 2t\langle \mathbf{S}_{\mathbf{m}'}^{\perp} \rangle \qquad (5.54)$$

where $\bar{v} = \frac{z}{2}\bar{v}_{\mathbf{mm}'}$, $t = zt_{\mathbf{mm}'}$, $\mathbf{m}' = \mathbf{m} + \mathbf{a}$, $\mathbf{e}$ is a unit vector in the z- direction, and $\mathbf{S}_{\mathbf{m}}^{\perp}$ is a spin component perpendicular to z. In the absence of the macroscopic current $\langle S^{y} \rangle = 0$ and at $T = 0$

$$\langle S_{\mathbf{m}}^{z} \rangle = \frac{1}{2}cos\Theta \qquad (5.55)$$

$$\langle S_m^x \rangle = \frac{1}{2} sin\Theta \qquad (5.56)$$

where Θ is the angle between z-axis and spin. The ground state has $\langle S_m \rangle$ parallel to H_m and we arrive at the following set of equations for Θ and μ

$$sin\Theta = \frac{tsin\Theta'}{\sqrt{(\mu + \bar{v}cos\Theta')^2 + t^2 sin^2\Theta'}}, \qquad (5.57)$$

$$cos\Theta = -\frac{\mu + \bar{v}cos\Theta'}{\sqrt{(\mu + \bar{v}cos\Theta')^2 + t^2 sin^2\Theta'}}, \qquad (5.58)$$

$$cos\Theta + cos\Theta' = 2(1 - 2n) \qquad (5.59)$$

where Θ' is the angle for the nearest neighbors.

Two solutions to Eq.(5.57-59) are possible. The first one is a 'ferromagnetic' solution

$$cos\Theta = cos\Theta' = 1 - 2n \qquad (5.60)$$

and

$$\mu = -(1 - 2n)(\bar{v} + t) \qquad (5.61)$$

In the 'ferromagnetic' state bipolarons are distributed uniformly over the lattice with the density per site n. The total energy of this state

$$\frac{E_f}{N} = -\frac{t}{4}\left(1 + (1 - 2n)^2(1 + \frac{\bar{v}}{t})\right) \qquad (5.62)$$

The second solution is an 'antiferromagnetic' one with two sublattices having different Θ and Θ'

$$cos\Theta = 1 - 2n + \sqrt{1 + (1 - 2n)^2 - \frac{2(1 - 2n)\bar{v}}{\sqrt{\bar{v}^2 - t^2}}} \qquad (5.63)$$

$$cos\Theta' = 1 - 2n - \sqrt{1 + (1 - 2n)^2 - \frac{2(1 - 2n)\bar{v}}{\sqrt{\bar{v}^2 - t^2}}} \qquad (5.64)$$

It exists if only $\bar{v} > t$ and the density is sufficiently high

$$n > n_c = \frac{1}{2}\left(1 - \sqrt{\frac{\bar{v} - t}{\bar{v} + t}}\right). \qquad (5.65)$$

In the region of its existence the 'antiferromagnetic' state is a ground state because

$$E_a = -\frac{\bar{v}N}{4} < E_f \qquad (5.66)$$

The conclusion is that the bipolarons exist at $T = 0$ in two states: as a homogeneous quantum liquid resembling He^4 or at high density as a mixture of an inhomogeneous Bose-Einstein condensate $(sin\Theta \neq sin\Theta' \neq 0)$ and a charge density wave $cos\Theta \neq cos\Theta'$. At a very low density the Wigner crystallization of charged bipolarons is feasible because of their long-range Coulomb repulsion.

One can expect that the excitation spectrum is similar to that of a Heisenberg magnet and one-particle excitations are 'magnons'. At $T = 0$ one can write down the equation of the 'spin' motion

$$\frac{d\mathbf{S_m}}{dt} = \mathbf{H_m} \times \mathbf{S_m} \qquad (5.67)$$

We allow each 'spin' besides its static component Eq.(5.55,56) to have a small time and space dependent part

$$\delta \mathbf{S} exp(i\mathbf{k} \cdot \mathbf{m} - i\omega t) \qquad (5.68)$$

From Eq.(5.67)

$$-i\omega \delta S^x = -t\delta S^y sin\Theta' cot\Theta + (t - E_\mathbf{k})\delta S'^y cos\Theta \qquad (5.69)$$

$$
\begin{aligned}
-i\omega \delta S^y &= t\delta S^x sin\Theta' cot\Theta - (t - E_\mathbf{k})\delta S'^x cos\Theta \\
&- t\delta S^z sin\Theta' - \frac{\bar{v}}{t}(t - E_\mathbf{k})\delta S'^z sin\Theta \qquad (5.70)
\end{aligned}
$$

$$-i\omega \delta S^z = t\delta S^y sin\Theta' - (t - E_\mathbf{k})\delta S'^y sin\Theta \qquad (5.71)$$

where $E_\mathbf{k} = \sum_{\mathbf{m'}} t_{\mathbf{mm'}}[1 - exp(i\mathbf{k} \cdot (\mathbf{m} - \mathbf{m'}))]$ is the energy dispersion of a single bipolaron on a lattice. From the condition of the existence of a nontrivial solution for Eq.(5.69-71) one obtains the excitation spectrum of the 'ferromagnetic' ground state $\omega = \epsilon_\mathbf{k}$:

$$\epsilon_\mathbf{k} = \sqrt{E_\mathbf{k}\left(t + [\bar{v} - (1 - 2n)^2(\bar{v} + t)](1 - \frac{E_\mathbf{k}}{t})\right)} \qquad (5.72)$$

with **k** varying in the first Brillouin zone (for intersite bipolarons in a simple lattice it is half of the original one). In the long-wave limit $k \to 0$ this spectrum is sound-like as in He^4, Fig.5.2f

$$\epsilon_\mathbf{k} = sk \tag{5.73}$$

with the 'sound' velocity

$$s = 2a\sqrt{\frac{t(\bar{v}+t)n(1-n)}{z}} \tag{5.74}$$

The linear dispersion is, of course, the consequence of the nearest neighbor approximation, used here. The long-range Coulomb interaction yields the plasma gap in three dimensions and the square root dispersion in $2d$ (Chapter 1).

When the density is critical $n = n_c$, Eq.(5.65) the spectrum is given by

$$\epsilon_\mathbf{k} = \sqrt{E_\mathbf{k}(2t - E_\mathbf{k})} \tag{5.75}$$

and the critical velocity $v_c = min\frac{\epsilon_\mathbf{k}}{k}$ is zero, Fig 5.2f.

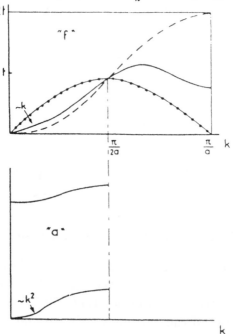

Fig.5.2. The excitation spectrum (solid lines) of BS (f) and M (a) ground states. Dashed line refers to a single bipolaron, dotted line to $n = n_c$.

Above n_c the charge density wave develops and the excitation spectrum consists of two branches

$$\epsilon_{\mathbf{k}}^{\pm} = \sqrt{\gamma^2 t^2 + E_{\mathbf{k}}(2t - E_{\mathbf{k}}) \pm \gamma t \sqrt{\gamma^2 t^2 + 2E_{\mathbf{k}}(2t - E_{\mathbf{k}})}} \quad (5.76)$$

with

$$\gamma^2 = 2\frac{\bar{v}^2 - t^2}{t^2}\left(1 + (1 - 2n)^2 - 2(1 - 2n)\frac{\bar{v}}{\sqrt{\bar{v}^2 - t^2}}\right) \quad (5.77)$$

and $\mathbf{k}$ varying in a new Brillouin zone, Fig. 5.2a. In the long-wave limit

$$\epsilon^+ = t\gamma\sqrt{2} \quad (5.78)$$

and

$$\epsilon^- \sim k^2 \quad (5.79)$$

The gap in the spectrum is of order of $\bar{v}$ if $\bar{v} \gg t$.

5.2.3 $T - n$ phase diagram

At finite temperatures thermal as well as quantum fluctuations are important. It is clear that under the condition $\bar{v} \gg t$ the off-diagonal long range order $(ODLRO)$ (i.e. the Bose-Einstein condensate) disappears first with increasing temperature at $T \sim t$, followed by the disappearance of the charge ordered state $(DLRO)$ at $T \sim \bar{v}$. At temperature $\bar{v} < T < T^{**}$ (Eq.(4.76)) there exists an unusual metal of nondegenerate bipolarons with an elementary charge $2e$. As a result the $T - n$ phase diagram of a bipolaronic liquid consists of four phases: two of them are low-temperature phases: a bipolaronic superfluid (BS) and a mixed phase (M) with $ODLRO$, described above and two high-temperature phases, one of them is an unusual metal (N) and the other is a charge ordered state (CO). In an extreme limit of a very high phonon frequency $\omega \gg \Delta$ the bipolaron bandwidth t and a short-range component of the repulsion $v^{(2)}$ are of the second order in the polaron bandwidth w (section 5.2.1) for on-site bipolarons. In this limit the on-site bipolaron hamiltonian can be mapped on the negative U Hubbard hamiltonian if the long-range Coulomb interaction is screened. In a more realistic case of long-range forces and (or) $\omega < \Delta$ no such mapping is possible. Nevertheless in a qualitative analysis of the phase diagram we can use the finite temperature mean-field (MFA) and random phase approximations (RPA), developed for the *negative U* Hubbard hamiltonian by Robaszkiewicz *et al.* (1981).

The MFA phase diagram is shown in Fig.5.3 for $\bar{v} = 2t$.

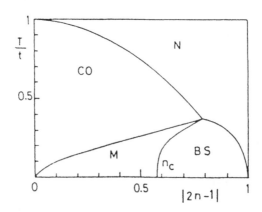

Fig.5.3. Mean-field $T - n$ phase diagram of bipolarons, $\bar{v}/t = 2$.

Quantum fluctuations ('magnons'), which can be taken into account with the RPA equation of motion for the 'spin-spin' correlation function lead to a significant modification of the critical density n_c. Quantum fluctuations extend the region of the existence of the BS phase which turns out to be the ground state $(T = 0)$ even in the limit $\bar{v}/t \to \infty$ if the density is low,

$$n < 0.078 \tag{5.80}$$

for a simple cubic lattice.

T_c is determined by (Alexandrov *et al.*(1986a)):

$$\frac{n}{1 - 2n} = \frac{1}{N} \sum_{\mathbf{k}} \left(exp\frac{(1 - 2n)E_{\mathbf{k}}}{T_c} - 1 \right)^{-1} \tag{5.81}$$

With the density of states in the bipolaronic band

$$N_b(E) = \frac{1}{N} \sum_{\mathbf{k}} \delta(E - E_{\mathbf{k}})$$

this equation takes the form of the condition for the Bose-Einstein condensation of the ideal Bose gas if $n \ll 1$

$$n = \int dE \frac{N_b(E)}{exp(E/T_c) - 1}. \tag{5.82}$$

At higher density a hard core correction appears, Fig.5.4,

$$T_c \simeq \frac{3.31 n^{2/3}}{m^{**} a^2} \left(1 - 0.54 n^{2/3}\right) \tag{5.83}$$

T_c is independent of the dynamical repulsion of bipolarons $\bar{v}$ which is an artifact of the random phase and nearest neighbor approximations.

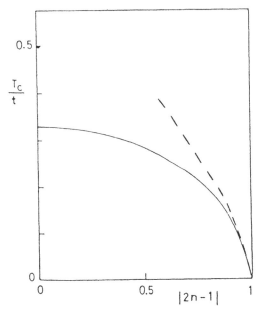

Fig.5.4. Hard-core correction (solid line) to the critical temperature of the ideal Bose-gas (dashed line).

Taking into account the next neighbor interaction Kubo and Takada (1983) found two new phases of bipolaronic liquid: an incommensurate CO phase and an incommensurate M phase. They argued that thermal and quantum fluctuations lead to a linear dispersion of low energy excitations of the commensurate M phase rather than the quadratic one, Eq.(5.79). If so the charge ordered low temperature phase M is also superfluid.

We believe, however that in real solids the long range Coulomb repulsion of bipolarons is the only relevant term for low-energy kinetics and thermodynamics. That favors the homogeneous charged Bose-liquid for all realistic values of the Coulomb interaction and density. A mapping of the bipolaronic hamiltonian on a charged Bose gas is useful in this case.

5.2.4 Mapping on a charged Bose-gas

One can transform the bipolaronic hamiltonian to a representation containing only Bose operators $a_{\mathbf{m}}$, $a_{\mathbf{m}}^{\dagger}$

$$b_{\mathbf{m}} = \sum_{k=0}^{\infty} \beta_k a_{\mathbf{m}}^{\dagger k} a_{\mathbf{m}}^{k+1} \tag{5.84}$$

$$b_{\mathbf{m}}^{\dagger} = \sum_{k=0}^{\infty} \beta_k a_{\mathbf{m}}^{\dagger k+1} a_{\mathbf{m}}^{k} \tag{5.85}$$

with

$$a_{\mathbf{m}} a_{\mathbf{m}'}^{\dagger} - a_{\mathbf{m}'}^{\dagger} a_{\mathbf{m}} = \delta_{\mathbf{m},\mathbf{m}'}. \tag{5.86}$$

The first few coefficients β_k are determined with the substitution of Eq.(5.84,85) into the commutation rules Eq.(5.48,49)

$$\beta_0 = 1, \beta_1 = -1, \beta_2 = \frac{1}{2} + \frac{\sqrt{3}}{6} \tag{5.87}$$

We introduce further the field bipolaron and boson operators

$$\phi(\mathbf{r}) = \frac{1}{\sqrt{N}} \sum_{\mathbf{m}} \delta(\mathbf{r} - \mathbf{m}) b_{\mathbf{m}} \tag{5.88}$$

$$\psi(\mathbf{r}) = \frac{1}{\sqrt{N}} \sum_{\mathbf{m}} \delta(\mathbf{r} - \mathbf{m}) a_{\mathbf{m}} \tag{5.89}$$

where $\delta(\mathbf{r} - \mathbf{m})$ is the eigenfunction of the coordinate operator. The transformation for the field operators takes the form

$$\phi(\mathbf{r}) = \left(1 - \frac{\psi^{\dagger}(\mathbf{r})\psi(\mathbf{r})}{N} + \frac{(1/2 + \sqrt{3}/6)\psi^{\dagger}(\mathbf{r})\psi(\mathbf{r})\psi(\mathbf{r})}{N^2} + ...\right)\psi(\mathbf{r}) \tag{5.90}$$

and the bipolaronic hamiltonian

$$H_b = -\int dr dr' \psi^\dagger(\mathbf{r}) \left(t(\mathbf{r} - \mathbf{r}') + \mu\right) \psi(\mathbf{r}') + H_{int} \qquad (5.91)$$

with

$$H_{int} = H_l + H_s, \qquad (5.92)$$

where H_l is the dynamic part of the interaction

$$H_l = \frac{1}{2}\int dr dr' \bar{v}(\mathbf{r} - \mathbf{r}')\psi^\dagger(\mathbf{r})\psi^\dagger(\mathbf{r}')\psi(\mathbf{r})\psi(\mathbf{r}'), \qquad (5.93)$$

and H_s describes the kinematic hard core effects

$$\begin{aligned} H_s &= \frac{2}{N}t(\mathbf{r} - \mathbf{r}')(\psi^\dagger(\mathbf{r})\psi^\dagger(\mathbf{r}')\psi(\mathbf{r}')\psi(\mathbf{r}') \\ &+ \psi^\dagger(\mathbf{r})\psi^\dagger(\mathbf{r})\psi(\mathbf{r})\psi(\mathbf{r}')) + H^{(3)}. \end{aligned} \qquad (5.94)$$

Here

$$t(\mathbf{r} - \mathbf{r}') = \sum_{\mathbf{k}}(t - E_{\mathbf{k}})e^{i\mathbf{k}\cdot(\mathbf{r} - \mathbf{r}')}, \qquad (5.95)$$

$$\bar{v}(\mathbf{r} - \mathbf{r}') = \frac{1}{N}\sum_{\mathbf{k}}\bar{v}_{\mathbf{k}}e^{i\mathbf{k}\cdot(\mathbf{r} - \mathbf{r}')}, \qquad (5.96)$$

and $\bar{v}_{\mathbf{k}} = \sum_{\mathbf{m}'\neq\mathbf{m}}\bar{v}_{\mathbf{m},\mathbf{m}'}exp(i\mathbf{k}\cdot\mathbf{m})$ is the Fourier component of the bipolaron repulsion. The term $H^{(3)}$ contains higher powers of the field operators than four. The essential physics of bipolarons is controlled by the two-particle interaction, which includes a short-range contribution $t(\mathbf{r} - \mathbf{r}')$ due to the hard-core effect. Because $\bar{v}$ contains the short range part $v^{(2)}$ this additional contribution can be included in the definition of $\bar{v}$. As a result H_b is identical to the hamiltonian of charged bosons tunneling in a band.

5.2.5 Bipolaron electrodynamics

To describe electrodynamics of bipolarons one can take into account the vector potential $\mathbf{A}(\mathbf{r})$ of the external field with the Peierls substitution (Peierls (1933))

$$t_{\mathbf{m},\mathbf{m}'} \rightarrow t_{\mathbf{m},\mathbf{m}'}e^{-i2e\mathbf{A}(\mathbf{m})\cdot(\mathbf{m}-\mathbf{m}')} \qquad (5.97)$$

which is a fair approximation if the magnetic field is weak compared with the atomic field

$$eHa^2 \ll 1. \qquad (5.98)$$

Here $\mathbf{A}(\mathbf{r})$ is a vector potential which can be also time dependent. This yields in real space

$$t(\mathbf{r}, \mathbf{r}') = \sum_{\mathbf{k}} (t - E_{\mathbf{k}+2e\mathbf{A}}) e^{i\mathbf{k}\cdot(\mathbf{r}-\mathbf{r}')}, \qquad (5.99)$$

If the condition Eq.(5.98) is satisfied one can expand $E_{\mathbf{k}}$ in the vicinity of $\mathbf{k} = 0$ to obtain

$$t(\mathbf{r}, \mathbf{r}') \simeq \left(t + \frac{(\nabla - 2ie\mathbf{A}(\mathbf{r}))^2}{2m^{**}}\right) \delta(\mathbf{r} - \mathbf{r}') \qquad (5.100)$$

where

$$\frac{1}{m^{**}} = \frac{d^2 E_{\mathbf{k}}}{d\mathbf{k}^2},$$

with $k \to 0$. As a result the hamiltonian of fermions strongly coupled with any bosonic field (i.e. phonons) reduces to

$$\begin{aligned} H_b &= -\int d\mathbf{r}\psi^\dagger(\mathbf{r}) \left(\frac{(\nabla - 2ie\mathbf{A}(\mathbf{r}))^2}{2m^{**}} + \mu\right)\psi(\mathbf{r}) \\ &+ \frac{1}{2}\int d\mathbf{r}d\mathbf{r}'\bar{v}(\mathbf{r} - \mathbf{r}')\psi^\dagger(\mathbf{r})\psi^\dagger(\mathbf{r}')\psi(\mathbf{r})\psi(\mathbf{r}') \quad (5.101) \end{aligned}$$

if three body and higher order interactions are neglected. The hard core effect is included in the definition of the repulsion $\bar{v}$ as described above, and the constant t in Eq.(5.100) is absorbed by the chemical potential μ.

The electrodynamics of the condensed charged Bose-liquid can be derived using the evolution of the Heisenberg field operator $\psi(\mathbf{r}, t)$

$$\frac{\partial \psi(\mathbf{r}, t)}{\partial t} = i[H_b, \psi(\mathbf{r}, t)],$$

which yields

$$\begin{aligned} i\frac{\partial \psi(\mathbf{r}, t)}{\partial t} &= -\left(\frac{(\nabla - 2ie\mathbf{A}(\mathbf{r}))^2}{2m^{**}} + \mu\right)\psi(\mathbf{r}, t) \\ &+ \int d\mathbf{r}'\bar{v}(\mathbf{r} - \mathbf{r}')\psi^\dagger(\mathbf{r}', t)\psi(\mathbf{r}', t)\psi(\mathbf{r}, t). \quad (5.102) \end{aligned}$$

We perform the Bogoliubov displacement transformation

$$\psi(\mathbf{r}, t) = \psi_0(\mathbf{r}, t) + \tilde{\psi}(\mathbf{r}, t) \qquad (5.103)$$

where $\psi_0(\mathbf{r}, t)$ is a c-number, describing the condensate, and $\tilde{\psi}(\mathbf{r}, t)$ is an excitation operator. Substitution of Eq.(5.103) in Eq.(5.102) yields an equation for the order parameter $\psi_0(\mathbf{r}, t)$

$$
\begin{aligned}
i\frac{\partial \psi_0(\mathbf{r}, t)}{\partial t} &= -\left(\frac{(\nabla - 2ie\mathbf{A}(\mathbf{r}))^2}{2m^{**}} + \mu\right)\psi_0(\mathbf{r}, t) \\
&+ \int d\mathbf{r}' \bar{v}(\mathbf{r} - \mathbf{r}')\left(n(\mathbf{r}', t) + \hat{V}\right)\psi_0(\mathbf{r}, t) \quad (5.104)
\end{aligned}
$$

with

$$
n(\mathbf{r}, t) = |\psi_0(\mathbf{r}, t)|^2 + \langle\tilde{\psi}^\dagger(\mathbf{r}, t)\tilde{\psi}(\mathbf{r}, t)\rangle \quad (5.105)
$$

the density of bosons and $\hat{V}$ the nonlocal interaction involving the fluctuations

$$
\begin{aligned}
\hat{V} &= \int d\mathbf{r}' \bar{v}(\mathbf{r} - \mathbf{r}')(\langle\tilde{\psi}(\mathbf{r}', t)\tilde{\psi}(\mathbf{r}, t)\rangle\psi_0^*(\mathbf{r}', t) \\
&+ \langle\tilde{\psi}^\dagger(\mathbf{r}', t)\tilde{\psi}(\mathbf{r}, t)\rangle\psi_0(\mathbf{r}', t))\psi_0^{-1}(\mathbf{r}, t).
\end{aligned}
$$

The equation (5.104) plays the same role as the Ginzburg -Landau (1950) equation for the BCS superconductor. We apply it in Chapter 7 to derive the upper critical field of the bipolaronic superconductor.

Fluctuations are described by the equation of motion for the field operator $\tilde{\psi}$, which in the Hartree approximation is

$$
\begin{aligned}
i\frac{\partial \tilde{\psi}(\mathbf{r}, t)}{\partial t} &= -\left(\frac{(\nabla - 2ie\mathbf{A}(\mathbf{r}))^2}{2m^{**}} + \mu\right)\tilde{\psi}(\mathbf{r}, t) \\
&+ \int d\mathbf{r}' \bar{v}(\mathbf{r} - \mathbf{r}')\left(n(\mathbf{r}', t) + \hat{U}\right)\tilde{\psi}(\mathbf{r}, t), \quad (5.106)
\end{aligned}
$$

with

$$
\begin{aligned}
\hat{U} &= \int d\mathbf{r}' \bar{v}(\mathbf{r} - \mathbf{r}')([\psi_0(\mathbf{r}', t)\psi_0(\mathbf{r}, t) + \langle\tilde{\psi}(\mathbf{r}', t)\tilde{\psi}(\mathbf{r}, t)\rangle]\tilde{\psi}^\dagger(\mathbf{r}', t) \\
&+ [\psi_0^*(\mathbf{r}', t)\psi_0(\mathbf{r}, t) + \langle\tilde{\psi}^\dagger(\mathbf{r}', t)\tilde{\psi}(\mathbf{r}, t)\rangle]\tilde{\psi}(\mathbf{r}', t))\tilde{\psi}^{-1}(\mathbf{r}, t).
\end{aligned}
$$

In the high density limit $r_s = 4m^{**}e^2/\epsilon_0(\frac{4\pi n}{3})^{1/3} \ll 1$ the number of bosons pushed up from the condensate by the repulsion is small as $r_s^{3/4}$, Eq.(1.31). Hence the contribution of the Hartree terms is small. Applying the linear transformation for $\tilde{\psi}$,

$$
\tilde{\psi}(\mathbf{r}, t) = \sum_\nu u_\nu(\mathbf{r}, t)\alpha_\nu + v_\nu^*(\mathbf{r}, t)\alpha_\nu^\dagger \quad (5.107)
$$

where α_ν is a bosonic quasiparticle annihilation operator in the state ν we arrive with the Bogoliubov-de Gennes type equations for the coefficients $u(\mathbf{r},t)$ and $v(\mathbf{r},t)$

$$
\begin{aligned}
i\frac{\partial u(\mathbf{r},t)}{\partial t} = &-\left(\frac{(\nabla - 2ie\mathbf{A}(\mathbf{r}))^2}{2m^{**}} + \mu\right)u(\mathbf{r},t) \\
&+ \int d\mathbf{r}'\bar{v}(\mathbf{r}-\mathbf{r}')(|\psi_0(\mathbf{r}',t)|^2 u(\mathbf{r},t) \\
&+ \psi_0^*(\mathbf{r}',t)\psi_0(\mathbf{r},t)u(\mathbf{r}',t) \\
&+ \psi_0(\mathbf{r}',t)\psi_0(\mathbf{r},t)v(\mathbf{r}',t))
\end{aligned}
\tag{5.108}
$$

and

$$
\begin{aligned}
-i\frac{\partial v(\mathbf{r},t)}{\partial t} = &-\left(\frac{(\nabla + 2ie\mathbf{A}(\mathbf{r}))^2}{2m^{**}} + \mu\right)v(\mathbf{r},t) \\
&+ \int d\mathbf{r}'\bar{v}(\mathbf{r}-\mathbf{r}')(|\psi_0(\mathbf{r}',t)|^2 v(\mathbf{r},t) \\
&+ \psi_0^*(\mathbf{r},t)\psi_0(\mathbf{r}',t)v(\mathbf{r}',t) \\
&+ \psi_0^*(\mathbf{r}',t)\psi_0^*(\mathbf{r},t)u(\mathbf{r}',t))
\end{aligned}
\tag{5.109}
$$

If $\mathbf{A} = 0$, states are classified with $\nu = \mathbf{k}$ and the coefficients are determined with Eq.(1.18-20). As a result the system of electrons strongly coupled with the lattice vibrations can be described as a charged Bose-liquid with $e^* = 2e$ and the effective boson mass m^{**} if the external field varies slowly in time and smoothly in space.

Chapter 6

Theoretical models of high-T_C oxides

6.1 Breakdown of the Fermi-liquid approach

Knowledge of the mechanism of pairing of carriers and of the nature of the normal state is central to an understanding of the high-T_c supeconductivity in metal oxides (Bednorz and Muller (1986)) and in doped fullerenes ($M_x C_{60}$) (Hebard *et al.* (1991), Holczer *et al.* (1991)). Superconductivity is a correlated many body state of pairs which is well described by the BCS theory in the weak-coupling limit $\lambda < 1$. Depending on the specific bosonic field, which 'glues' two carriers together, the BCS superconductor could be not only 'phononic' but also 'excitonic' (Little (1964), Ginzburg (1968)), 'plasmonic' (Fröhlich (1968), Pashitskii (1968)), or 'magnonic (quasi)' (Schrieffer *et al.* (1989), Millis *et al.* (1990), Wood and Cooke (1992), Kamimura *et al* (1992), Monthoux and Pines (1992)) (see left hand side of Fig. 6.1).

The BCS theory like any mean-field theory is rather universal, describing the cooperative-quantum phenomenon of superfluidity in He^3 with T_c a few mK, in a great number of superconducting metals and their alloys, and, as is believed, even in atomic nuclei with $T_c = 10^{10} K$ (Anderson and Schrieffer (1991)). One can ask: are metal oxides and doped fullerene exceptions and if so, why? Only experiment can give the answer. There are several hallmark indicators of the BCS superconductivity which can be checked experimentally like the isotope effect, the coherent peak in the nuclear spin relaxation rate, the exponential decrease be-

low T_c of the thermal conductivity and of the sound attenuation
(Chapter 2). The explicit temperature (or frequency) dependence
of different response functions is not so crucial. As an example,
the power law rather than exponential temperature behavior of
different thermodynamic or kinetic characteristics like electronic
heat capacity, the London penetration depth, thermal conductivity, sound attenuation are also compatible with the BCS theory if
the pairing occurs with a non-zero orbital momentum and the gap
$\Delta = 0$ for some parts of the Fermi-surface.

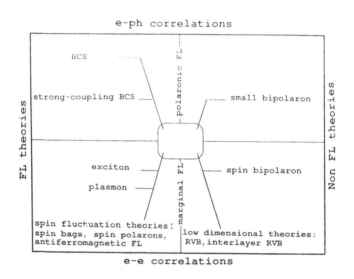

Fig. 6.1. BSC-like theories of high T_c with electron-phonon (top) and
electron-electron (bottom) interactions (left hand side). Right hand side:
non Fermi liquid theories with electron-phonon (top) and electron-electron
(bottom) correlations.

More crucial is the fact realized now that many experiments
show just the opposite behavior to that predicted by the BCS theory for the superconducting state including its canonical strong-
coupling and a non-zero orbital momentum generalizations, Table 6.1. The Fermi liquid model for the normal state is also in
clear contradiction with the low-frequency kinetics of copper-based
high-T_c oxides.

Thus the answer to the first part of the above question is yes,
the metal oxides are exceptions. Moreover it is not at all sur-

prising because the estimated Fermi-energy (if it exists) in metal oxides and in doped fullerene is unusually low, in the range of $0.1eV$ due to the low carrier density and heavy masses of carriers. The adiabatic approximation $\omega \ll E_F$, which makes the mean-field treatment of the electron-phonon (or any boson) interaction selfconsistent no longer holds. Both λ and μ are large, and the *logarithm* in the definition of the Coulomb pseudopotential μ^* does not apply. The carriers are strongly correlated in the normal and superconducting states both due to the Coulomb and the electron-phonon interactions.

However the absence of a small parameter does not mean that the mean-field scheme is unacceptable. The transition metals are a few of many examples of Fermi-liquids, in which the kinetic energy is less than the bare Coulomb energy. Nevertheless the Hartree-Fock and BCS treatment seems to describe adequately their low-frequency and low-temperature properties. Hence one can try to overcome the qualitative discrepancies with experiment, using the Eliashberg equations and making allowance for the pair-breaking effects due to a rather high critical temperature (Allen and Rainer (1991), Mikhailovsky *et al.* (1991), Carbotte (1990)), for antiferromagnetic fluctuations (Monthoux and Pines (1992)), and for the Van-Hove singularities of the electron density of states (Friedel (1989), Newns *et al.* (1992)). In this way one can suppress the Hebel-Slichter peak in NMR, describe the temperature dependence of some low-frequency dynamic properties , and the unusual variations of the isotope effect in several metal oxides. To be self-consistent these improvements of the BCS description require as a rule a large value of λ (> 0.5) and a combination of the phonon and non-phonon (e.g. due to spin fluctuations) interactions.

A description of normal state properties and the internal inconsistency (due to the polaron collapse) in the strong-coupling limit $\lambda > 1$ pose two serious problems within the framework of this approach. For example, the linear resistivity in the temperature range 10-1000K, the temperature dependent Hall and Korringa ratios, the thermoelectric power, decreasing with increasing temperature, as well as the infrared conductivity remain to be explained within the Fermi liquid approach.

Pines and co-workers argue that the magnetic interaction between planar quasiparticles in the cuprate superconductors leads

Table 6.1: BSC/FL predictions and experimental results for highT_c copper oxides

Physical property	BCS/FL prediction	experiment
$T < T_c$		
NMR $1/T_1$	coherent peak	absent
Isotope effect	$\alpha = 0.5$ or less	$-0.013 < \alpha < +1.0$
Thermal conductivity	decrease	enhancement
Gap	$3.5T_c$	$7 - 8T_c$ (for $T_c = 90K$)
$T > T_c$		
Hall ratio	constant	$\sim 1/T$
Korringa ratio $1/T_1T$	constant	temperature dependent
Thermopower	$\sim T$, small	nonlinear, large
Infrared conductivity	Drude-law	mid-infrared maxima

to a new quantum state of matter, the nearly antiferromagnetic Fermi liquid. It possesses a well-defined Fermi surface, but it is not the ordinary Landau Fermi liquid, because none of its magnetic or transport properties are those of Landau theory. With the soft antiferromagnetic fluctuations they explain a resistivity generally linear in T, a non-Drude like optical conductivity, and a non-Korringa behavior of NMR. However their model clearly contradicts the neutron data by Bruckel *et al* (1987), obtained on the high flux reactor at Grenoble using 32 detectors for 21 beam days to look for the diffuse magnetic scattering, which was not observed in the metallic phase of $YBa_2Cu_3O_{7-\delta}$ in the energy range up to $30meV$. It follows from these data that the quasi-elastic peak in neutron scattering and hence local magnetic moments are practically absent from the metallic phase.

There is a great deal of interest in attributing high T_c of the cuprates to the proximity of the Fermi level to a Van Hove singularity in the density of states. However Radtke *et al.* (1993) based on the Eliashberg equations conclude that this model constrained by neutron-scattering and transport measurements yields $T_c \simeq 10K$ for both $La_{2-x}Sr_xCuO_4$ and $YBa_2Cu_3O_{7-\delta}$. The Van Hove singularity enhances T_c much less effectively than weak-coupling *adhoc* calculations would suggest.

6.2 Non Fermi-liquid alternatives

The interesting alternative to the usual Fermi-liquid description is the proposals based on the low-dimensionality (2*d* rather than 3*d*) of copper-based high-T_c oxides, in particular Anderson's resonating-valence-bond (RVB) theory (Anderson (1987) and the interlayer RVB model by Wheatley, Hsu and Anderson (1988). The ground state in these models supports 'topological solitons' (spinons and holons in Anderson's terminology), such as occur in one dimensional models like the one dimensional Hubbard model. The main idea is that an electron injected into a layer decays into a singlet charge e component (holon) and a spin half component (spinon) and conversely, must form again in order to come out. This is the case for one dimensional correlated electrons, which form a so-called Luttinger liquid.

A Bose-quasiparticle implies a condensate. However there is no experimental evidence for a charge e superfluid. It was suggested therefore that the superconductivity of copper based high-T_c materials is due to the Josephson-like coupling of holons between the layers and the dominant process at sufficiently low temperatures is a coherent interlayer tunneling of holon pairs with charge 2e. Anderson argues that the existence of an upper Hubbard band would necessarily lead to the Luttinger liquid, as opposed to the Fermi liquid even in two dimensions. The spinon-holon phenomenology introduced by Anderson gives an appealing explanation of the linear in-plane resistivity, the strange shape and sharp cusp feature of the observed angle-resolved photoemission and several other experimental observations. There is no single-particle coherent tunneling between two spinon-holon planes above T_c. However there is a coherent two-particle tunneling between them below T_c. The corresponding kinetic energy in the hamiltonian is responsible for the BCS-like pairing at high temperature $T_c \simeq t_\perp^2/t$ and for the plasma-like gap, observed in the c-axis conductivity (see Chapter 7).

In our view the real trouble with the Anderson-Luttinger liquid is that for it to be valid it requires an absence of the diffractive scattering between holes of the cuprate superconductors. A known case where diffractive scattering is absent is that of one-dimensional interacting fermion systems. But in one-dimensional interacting fermion systems the scattering is nondiffractive due to topological reasons. In two - and higher -dimensional systems one does not have such topological constraints (see Lal and Joshi

(1992)).

Moreover measurements of the out -of-plane resistivity ρ_c of *BSCCO* (Xiang *et al.* (1992)) and of *YBCO* (Forro *et al.* (1992) show the metallic -like linear temperature dependence of ρ_c and the mean-free path in *c*-direction comparable with the interplane spacing. It seems that a well-developed interlayer band can eliminate low-dimensional hypotheses, which definitely fail to account for $d\rho_c/dT > 0$ of highly oxygenated samples. As we show in Chapter 7 many kinetic properties of copper oxides including *c*-axis transport can be explained with (bi)polarons.

Of course it is also conceivable (and even probable) that some high-T_c materials are *marginal* Fermi-liquids in the sense that the energy interval near the Fermi level in which the Landau quasiparticles exist is small compared with T_c. The heavy polaronic Fermi-liquid described in Section 5.1 is an example. Varma *et al.* (1989) proposed a phenomenological description of marginal Fermi-liquids based on the postulated properties of the response function. However it was shown by Mitani and Kurihara (1992) that Varma's assumption does not break the Fermi-liquid picture, as far as the low-energy excitations are concerned.

The Fermi-liquid behavior is destroyed by the strong Coulomb repulsion in a narrow band giving Mott insulator and by the strong electron-phonon and (or) electron-spin fluctuation interactions. We believe that the ground state of strongly coupled carriers in a doped Mott insulator is the bipolaronic charged Bose-liquid. Mott (1987), Alexandrov (1987), Prelovsek, Rice and Zhang (1987) and other authors proposed bipolarons as a key element for the understanding of the high-T_c phenomenon in metal oxides. The key point for the bipolaronic mechanism of high-T_c is the possibility of the coherent tunneling of (bi)polarons with a reasonable value of the effective mass. Several authors assert that their bandwidth should not exceed $10^{-4} - 10^{-5} eV$, and then the maximal T_c attainable with small bipolarons should be a few K or less. However, this is not the case for the intermediate value of the coupling constant $\lambda \simeq 1$ and (or) the high-frequency phonons $\omega \simeq 0.1 eV$. Because the optical phonon frequencies in metal oxides are high ($0.05 - 0.1 eV$) the polaron binding energy is large ($E_p = 0.2 - 0.4 eV$) and thus the value of the bare electronic bandwidth compatible with the small polaron formation is large enough being of the order of $1.0eV$. In this case the estimate of the small polaron bandwidth (Section 4.6)) yields the value of w a few hundred K and the same for T_c (Section 5.1.2). The fluctuations of the hopping integral due to the ion displacements (Ray (1987)) and correlation effects, destroying the electron-hole symmetry (Hirsch (1993)) reduce the polaron mass significantly. Of course, a short-range *intersite* Coulomb repulsion should be suppressed below $2E_p$ to ensure the formation of mobile intersite bipolarons, and that is quite feasible because the high-frequency dielectric constant in metal oxides is usually large, $\epsilon \sim 5$ and larger. The low-frequency dielectric constant is extremely large in oxides, normally $\epsilon_0 > 100$.

Our intention for the rest part of the book is to apply the bipolaron theory of strong-coupling superconductivity to high-T_c superconductors, in particular to $YBa_2Cu_3O_{7-\delta}$ for which the extensive experimental data are available. We show that our extension of the BCS theory to the strong-coupling limit (Chapter 5) gives a quantitative explanation of many features of low-frequency spin and charge kinetics of high- T_c oxides both in the normal and superconducting state. Phenomena treated include spin gap effects in the normal state NMR and neutron scattering, resistivity, the Hall effect, infrared conductivity, thermopower, c-axis transport and normal state heat capacity. We discuss the isotope effect, λ-like heat capacity, heat transport below and above T_c, the London penetration depth, critical magnetic field, coherence length, the symmetry of the order parameter, gap above T_c, the results obtained from the angle resolved photoemission and the metal-insulator transition.

As far as doped fullerene is concerned it is too early to make a definite conclusion with the experimental results now at hand. Nevertheless doped fullerene seems to be prepared by nature to be (bi)polaronic because of its *bare* nonadiabaticity. The phonon frequencies are high, $\omega \simeq 0.2eV$ and the bare Fermi energy is very low $E_F \simeq 0.1 - 0.2eV$. The *logarithm* in the definition of the Coulomb pseudopotential μ^* does not apply in this nonadiabatic case and the electron-phonon coupling should be strong $(\lambda > 1)$ to overcome the Coulomb repulsion. According to Chapter 4 this implies small polarons. The cluster structure of C_{60} favors bipolarons. Carriers coupled with the high frequency molecular vibrations could be bound in pairs within one C_{60} molecule (ball) if twice polaronic level shift exceeds the Coulomb repulsion. The latter is suppressed because of the large diameter of the ball and the static dielectric constant.

Chapter 7

Bipolaron theory of high-T_C oxides

7.1 Basic model of CuO_2 plane

The solution for two electrons interacting with a polarisable continuum shows that the bipolaron forms more easily in two dimensions than in three, and that the mean value of the pair radius is a few Angstroms for the values of the dielectric constants characteristic for copper oxides (Emin(1989), Verbist *et al.* (1991), Bassani *et al.* (1991)). These results are confirmed by the solution of a two-particle Schrödinger equation for the CuO_2-plane with two different types of short-range correlations (repulsion on copper and attraction on oxygen), which yields the radius of the bound state less then two lattice constants in the relevant region of correlations (Alexandrov and Kornilovich (1993)). Different types of possible polaron pairing in La_2CuO_4 were investigated by Zhang and Catlow (1991) by computer simulation techniques based on lattice energy minimisation. The binding energy of the pair was found to be strongly related to the pair distance and geometry of the pair. The nearest neighbor pairing is energetically favorable with the binding energy $0.119eV$ for $O^- - O^-$ bipolaron. Taking this into account we proposed a simple model for the copper-oxygen plane, which is a key structural element of all copper-based high-T_c oxides (Alexandrov and Mott (1993b)).

Our assumption is that *all electrons* are bound in small singlet or triplet *intersite* bipolarons stabilised by the lattice and spin distortion. Because the undoped plane has a half-filled $Cu3d^9$ band there is no space for bipolarons to move if they are intersite. Their Brillouin zone is half of the original electron one and completely filled with hard-core bosons. *Hole* pairs, which appear with doping, have enough space to move and they are responsible for the low-energy charge excitations of the CuO_2 plane. Above T_c a material such as $YBCO$ contains a non-degenerate gas of these hole bipolarons in singlet, or in triplet states. Triplets are separated from singlets by the spin gap J and have a slightly lower mass due to the lower binding energy, Fig.7.1,2.

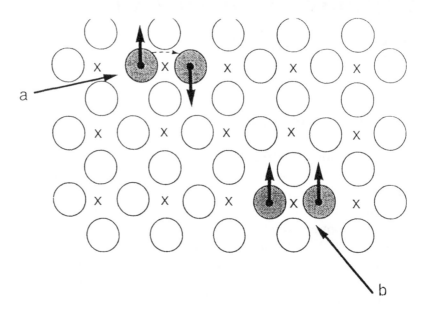

Fig.7.1.Inter-site singlet (a) and triplet (b) bipolarons on oxygen sites (circles) surrounded by copper (crosses). The internal optical transition of a singlet is shown by a dashed line.

The main part of the electron-electron correlation energy (Hubbard U) and the electron-phonon interaction is included in the binding energy of bipolarons and in their band-width renormalisation as described in Section V. The rest, including the boson-boson repulsion, is considered as the perturbation for extended hole states resulting in the canonical Boltzmann kinetics or in the Bogoliubov excitations in the superconducting state. In the normal state corrections due to the interaction to the single-particle spectrum are small if the RPA-parameter r_s for the Coulomb forces is not very large. To describe kinetic properties including the metal-insulator transition and Hall effect one should also take into account the Anderson localisation of bipolarons by disorder. This model has its obvious limitations as far as spin excitations are concerned. If all electrons are bound in small intersite bipolarons there are no local magnetic moments at $T = 0$. At finite temperatures some moments appear because singlets are thermally excited into triplet states. The neutron data by Bruckel *et al.* (1989) confirm the model. It follows from these data that the local magnetic moments are absent from the metallic phase. Many ESR studies show the same. The neutron data by Bruckel *et al.* have been quoted against the hypothesis that carriers are spin polarons. One of us (Mott (1993a)) has suggested that energy is gained with doping if spin polaron transforms into a metallic two dimensional sheet. On the other hand the insulating parent compounds are antiferromagnets as a rule. For a dense system of intersite singlet bipolarons the exchange interaction between them is important and could be responsible for the antiferromagnetic ground state of undoped materials. Approaching the metal-insulator transition the antiferromagnetic fluctuations dominate the electron-phonon interaction. As a result holes become Zhang-Rice singlets or spin polarons (section 4.8). Therefore we expect that the spin gap observed with NMR, neutron scattering, resistivity and heat capacity in copper based high-T_c oxides depends on the doping in accordance with several experimental observations (see Section 7.2).

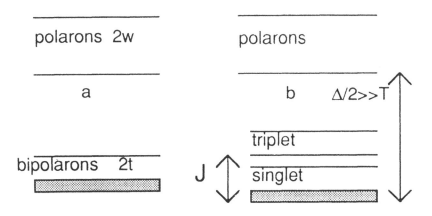

Fig.7.2. On-site (*a*) and inter-site (*b*) bipolaronic energy bands.

Keeping these limitations in mind we suggest that in the metallic region of doping the low-energy band structure consists of two bosonic bands (singlet and triplet), separated by the singlet-triplet exchange energy J, estimated to be of the order of a few tens meV, Fig.7.2. The half-bandwidth t is of the same order. The bipolaron binding energy is assumed to be large, $\Delta \gg T$ and therefore single polarons are irrelevant in the temperature region under consideration except for overdoping. This bandstructure is applied for hole pairs. We argue that many features of spin and charge excitations of metal oxides can be described within our simple model, Fig.7.2.

The population of singlet and triplet bands is controlled by the chemical potential $\mu = Tlny$, where y is determined from the thermal equilibrium of singlet and triplet bipolarons; if $T > T_c$

$$\ln\left(\frac{1 - ye^{-2t/T}}{1 - y}\right) + 3\ln\left(\frac{1 - ye^{(-2t-J)/T}}{1 - ye^{-J/T}}\right) = \frac{2nt}{T}, \qquad (7.1)$$

and $y = 1$ if $T < T_c$. Here n is the total number of pairs per cell and the density of states is assumed to be energy independent within bands.

7.2 Normal state

7.2.1 Spin gap in NMR and neutron scattering

Taking the usual contact hyperfine coupling of nuclei with electron spins, Eq.(4.78) and performing transformation to polarons and bipolarons as described in Chapter 5 one obtains the effective interaction of triplet bipolarons with nuclear spins (Alexandrov (1992c))

$$H_{int} \sim \sum_{\mathbf{m},l,l'} b^\dagger_{\mathbf{m},l} b_{\mathbf{m},l'} + h.c. \qquad (7.2)$$

where only intra-cell terms $\mathbf{m} = \mathbf{m}'$ are left. The NMR width due to the spin-flip scattering of triplet bipolarons on nuclei is obtained with the Fermi-golden rule:

$$\frac{1}{T_1} \sim \int d\xi N_t^2(\xi) f(\xi) (1 + f(\xi)) \qquad (7.3)$$

with $f(\xi) = (exp\xi/T - 1)^{-1}$ and $N_t(\xi)$ is the DOS in the triplet band. With $N_t \simeq 1/2t$ one obtains

$$\frac{1}{T_1} \simeq \frac{AT sinh(t/T)}{t^2[cosh((t+J)/T - lny) - cosh(t/T)]} \qquad (7.4)$$

where A a temperature independent hyperfine coupling constant being of the same order as in simple metals.

With Eq.(7.4) one can understand the main features of the nuclear spin relaxation rate in copper-based oxides: the absence of the Hebel-Slichter coherent peak below and the temperature dependent Korringa ratio $1/TT_1$ above T_c, the large value of $1/T_1$ due to the small bandwidth t, and fit the experimental data with reasonable values of the parameters, t and J, Fig.7.3.

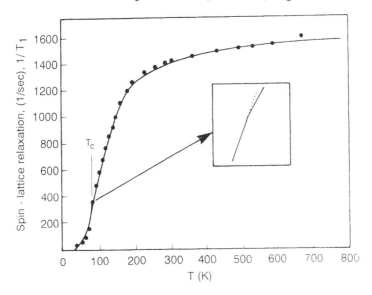

Fig.7.3. The temperature dependence of the nuclear spin relaxation rate $1/T_1$ (Machi *et al.* (1991)) compared with the theory ($J = 15meV$) showing the absence of the Hebel-Slichter peak of Cu NMR in $YBa_2Cu_4O_8$.

We compare the formula (7.4) with the representative experimental data by Machi et al (1991) for a wide temperature range in $YBa_2Cu_4O_8$. Similar unusual behavior of NMR was found in underdoped $YBa_2Cu_3O_{7-\delta}$. Of course, the fitting parameters t and J vary with the chemical formula. The Knight shift, which measures the spin susceptibility of carriers drops well above T_c in many copper oxides. This confirms our interpretation of the temperature dependent magnetic susceptibility as due to the triplet bipolarons with the temperature dependent density.

The value of the singlet-triplet bipolaron exchange energy J, determined from the fit to the experimental NMR width is close to the 'spin' gap observed above and below T_c in the underdoped *YBCO* with unpolarised (Rossat -Mignod *et al.* (1992)), Fig.7.4 and polarised (Mook *et al.* (1993)) neutron scattering, which reflects both having the same origin. The disappearance of the magnetic scattering for the energy transfer above approximately $50meV$ (Fig.7.4) is due to the narrowness of bipolaron bands, as is shown on Fig.7.2. For the energy transfer above $2t+J$ there are no final states available. The characteristic q-width of the magnetic susceptibility, determined from the neutron scattering, is comparable with the reciprocal lattice vector. This fact demonstrates the local character of magnetic excitations of the CuO_2 plane, which is the characteristic feature of triplet small bipolarons.

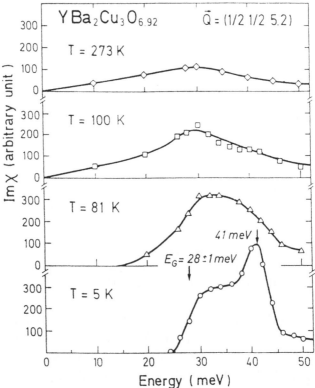

Fig.7.4. Spin gap in the spin fluctuation spectrum and the disappearance of the magnetic scattering above $50meV$ (Rossat-Mignod *et al.* (1993)).

7.2.2 Anderson localisation of bipolarons

To describe kinetic properties of bipolarons in copper oxides one should take into account their localization in a random potential. Because of low dimensionality ($2d$ rather then $3d$) any random potential leads to localization no matter how weak it is. Coulomb repulsion limits the number of bosons in each localised state, so that the distribution function will show a mobility edge E_c (Mott (1993a)).

The density of extended (free) bosons is given by (for simplicity we do not distinguish here singlets and triplets):

$$n_b(T) = \frac{T}{2t} \ln \left(\frac{1 - y e^{-2t/T}}{1 - y} \right) \qquad (7.5)$$

The intuitive picture of interacting bosons with a short-range interaction filling up all localized single-particle states in a random potential and Bose-condensing into the first extended state is known in the literature (see Hertz *et al.* (1979), Ma *et al.* (1986) and Fisher *et al.* (1989)). To calculate the density of localised bosons $n_L(T)$ one should take into account the repulsion between them. One can not ignore the fact that the localization length ξ generally varies with energy and diverges at the mobility edge. One would expect that the number of *hard core* bosons in a localized state near the mobility edge diverges in a similar way as the localization length does. However in our case of *charged* bosons their number in a single potential well is determined by the competition between their long-range Coulomb repulsion $\simeq 4e^2/\xi$ and the binding energy $E_c - \epsilon$. If the localization length diverges with the critical exponent $\nu < 1$: $(\xi \sim (E_c - \epsilon)^{-\nu})$, one can apply a 'single well-single particle' approximation assuming that one can place only one boson in each potential well. In doped semiconductors the exponent ν depends on the degree of compensation varying from $\nu = 0.5$ in SiP to $\nu \simeq 1.0$ in amorphous $NbSi$. In an extreme case of the hydrogen atom the average electron-nucleus distance is proportional to the inverse binding energy , i.e. $\nu = 1$. In this case the hydrogen negative ion exists but with rather low electron affinity (see Massey (1976)) so the doubly charged negative ion H^{2-} exists only as a resonant state. The gross features of the temperature behavior of $n_L(T)$ are not influenced by this approximation if the number of bosons in a potential well is finite.

Within this approximation localized *charged* bosons obey the Fermi-Dirac statistics:

$$n_L(T) = \int_{-\infty}^{E_c} \frac{N_L(\epsilon)d\epsilon}{exp(\frac{\epsilon-\mu}{T})+1}, \qquad (7.6)$$

where $N_L(\epsilon)$ is the density of localized states. Near the mobility edge it remains constant $N_L(\epsilon) \simeq \frac{n_L}{\gamma}$ with γ of order of a binding energy in a single random potential well and n_L the total number of localized states per unit cell. The integral in Eq.7.6) yields:

$$n_L(T) = n_L - n_L \sum_{k=1}^{\infty} (-1)^{k-1} \frac{y^k}{1+\frac{\gamma k}{T}} \qquad (7.7)$$

We chose the position of the mobility edge as zero, $E_c = 0$. The second term in Eq.(7.7), which is the number of empty localised states, turns out to be linear as a function of temperature in a wide temperature range $T < 2w, \gamma$ because the chemical potential is pinned in this temperature region to the mobility edge, $\mu \simeq E_c = 0$. This follows from the conservation of the total number of bosons per cell, $n = n_b(T) + n_L(T)$, which gives for the chemical potential:

$$\frac{T}{2w}ln\frac{1}{1-y} - \frac{n_L T}{\gamma}ln(1+y^{-1}) = n - n_L. \qquad (7.8)$$

If $T << (\gamma, 2w)$, the solution of this equation is $y \simeq 1$ with an exception of a very narrow region of concentration $n - n_L << T/2w$, where y decreases to about 0.6. The density n_b depends on y logarithmically, therefore its temperature dependence remains practically linear up to $T \simeq \gamma$:

$$n_b(T) = n - n_L + n_L bT, \qquad (7.9)$$

with temperature independent $b = ln(1 + y)/\gamma$.

In a more general case the statistics of localised bipolarons are different from Fermi and Bose-Einstein statistics. That could be important for some kinetic and thermodynamic properties like the thermoelectric power and specific heat.

7.2.3 The Hall effect and resistivity

Solving the Boltzmann equation with a weak magnetic field for extended bosons, scattered by acoustical phonons, by each other and by unscreened random potential one obtains the canonical expressions for the Hall ratio R_H, resistivity ρ and $cot\Theta_H = \rho/R_H$ (Alexandrov et al (1994)):

$$R_H = \frac{v_0 < \tau^2 >}{2en_b(T) < \tau >^2},\qquad(7.10)$$

$$\rho = \frac{v_0 m^{**}}{4e^2 n_b(T) < \tau >},\qquad(7.11)$$

$$cot\Theta_H = \frac{< \tau >}{\omega_c < \tau^2 >},\qquad(7.12)$$

where m^{**} is the in-plane boson mass, $\omega_c = 2eH/m^{**}$, τ the transport relaxation time, v_0 the volume of an elementary cell $(0.167nm^3$ for $YBa_2Cu_3O_7)$, and $< ... >$ means an average with energy and the derivative of the Bose-Einstein distribution function.

The transport relaxation rate due to the two-dimensional boson-phonon scattering has been shown to be energy independent and linear in temperature (Section 7.3.3):

$$\frac{1}{\tau_{b-ac}} = m^{**}C_{ac}T \qquad(7.13)$$

where the constant C_{ac} is proportional to the deformation potential.

In case of Bose-Einstein statistics umklapp scattering can be neglected, so the scattering between bosons in extended states does not contribute to the resistivity. However the inelastic scattering of an extended boson by localised bosons makes a contribution because the momentum is not conserved in two-particle collisions in the presence of the impurity potential. In a 'single well-single particle approximation' the role of the Pauli exclusion principle is played by the dynamical repulsion between bosons. That is why the boson-boson relaxation rate has the same temperature dependence as the fermion-fermion scattering and identical to that calculated by Xing and Liu (1991) in case of localised fermions.

The relaxation rate is proportional to temperature squared because only localized bosons within the energy shell of the order of T near the mobility edge contribute to the scattering and because the number of the final states is proportional to temperature:

$$\frac{1}{\tau_{b-b}} = \frac{\alpha e^2 b n_L}{m^{**}} T^2, \tag{7.14}$$

with α a constant.

Unoccupied impurity wells with density $b n_L T$ also contribute to the scattering giving rise to the energy independent elastic relaxation rate, which is linear in temperature:

$$\frac{1}{\tau_{b-im}} = m^{**} C_{im} n_L T \tag{7.15}$$

with C_{im} as a constant.

Substitution of Eq.(7.13-15) into Eq.(7.10-12) yields:

$$R_H = \frac{v_0}{2e(n - n_L + b n_L T)}, \tag{7.16}$$

$$\rho = ((m^{**})^2 C v_0 / 4 e^2) \frac{T + \sigma_b T^2}{n - n_L + b n_L T} \tag{7.17}$$

with $C = C_{ac} + n_L C_{im}$ and $\sigma_b = \alpha e^2 b n_L / (m^{**})^2 C$ the elastic and boson-boson scattering cross-sections,

$$cot\Theta_H = \frac{(m^{**})^2 C}{2eH} \left(T + \sigma_b T^2 \right) \tag{7.18}$$

These formulas contain important information about the number of bosons, localized states, and the relative strength of different scattering channels. There are two fitting parameters, n and n_L, if no significant variation of $(m^{**})^2 C, b, \sigma_b$ is expected with doping. Because of the presence of chains the number of in-plane carriers is not fixed by the chemical formula at least in $YBa_2Cu_3O_{7-\delta}$.

Carrington et al. (1993) and Ito et al. (1993) are presented in-plane kinetic data for single homogeneous crystals of $YBa_2Cu_3O_{7-\delta}$ in a wide range of doping and temperatures. The theoretical fit with parameters n and n_L is shown on Fig.7.5-7. These parameters as well as others are presented in Table 7.1..

One can see from the Table that the density of localized states increases with doping . The number of extended bosons increases

Table 7.1: Microscopic parameters determined from the Hall effect and resistivity of $YBa_2Cu_3O_{7-\delta}$.

δ	$n - n_L$	$bn_L(\times 10^{-3}K^{-1})$	$(m^{**})^2C(\times 10^{-18}\frac{kg}{sec \cdot K})$	$\sigma_b(\times 10^{-2}K^{-1})$
0.05	0.103	2.22	0.81	1.0
0.19	0.041	1.61	0.69	1.1
0.23	0.035	1.25	0.62	1.2
0.28	0.023	1.11	0.46	1.6
0.39	0.007	0.85	0.46	1.6

and the scattering cross-sections and their relative contribution depend slightly on the doping. Underdoped samples ($\delta > 0.2$) are practically compensated (the total number of bosons n is very close to the number of localised states n_L). This should be the case if every additional oxygen ion gives a single localised state.

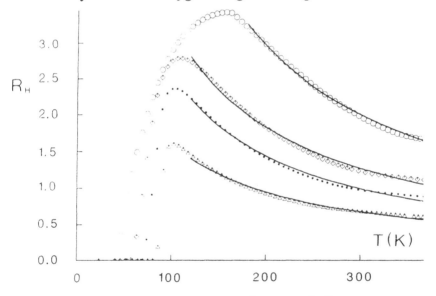

Fig.7.5. The Hall coefficient (in units $10^{-9}coulomb^{-1}m^3$) for $YBa_2Cu_3O_{7-\delta}$ compared with the theory (solid lines) for different δ: 0.05 ($\triangle$); 0.19 ($\bullet$); 0.23 ($\diamond$) and 0.39 (o). Parameters of the model are presented in Table 7.1.

The boson-phonon scattering is mainly responsible for the linear temperature dependence of ρ at low temperatures while the boson-boson scattering and the temperature dependent concentration $n_b(T)$ are responsible for the linear ρ at higher temperatures. *The residual resistivity is taken to be zero.* The slope of $cot\Theta_H$ at low temperature increases with the impurity concentration n_L

due to the elastic scattering by the random unoccupied potential wells. Any *nondegenerate* carriers on a two-dimensional lattice have the same temperature and doping dependencies of their kinetic properties, Eq.(7.16-18). This is also the case for triplet bipolarons. However if triplets are lighter then singlets the slope of the resistivity (proportional to $(m^{**})^2C$) diminishes with temperature because singlets are thermally excited into triplets. The characteristic small deviation from linearity (see Fig.7.6) should appear at temperature T^* where a 'spin' gap in NMR and in neutron scattering appears (Section 7.2.1). The change in the resistivity slope at $T = T^*$ was measured by Bucher *et al.* (1993) in $YBa_2Cu_4O_8$ and explained by Mott (1993b). Ito et al (1993) observed a correlation in the deviation of resistivity from the linear one with the temperature dependent Korringa ratio in $YBa_2Cu_3O_{7-\delta}$. These observations support our explanation of the spin gap as the singlet-triplet bipolaron exchange energy.

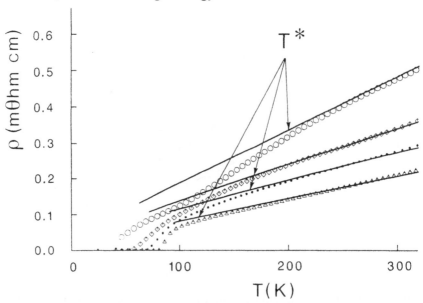

Fig.7.6. Resistivity for $YBa_2Cu_3O_{7-\delta}$ compared with the theory.

7.2.4   Thermoelectric power

The thermoelectric power S of high-T_c oxides shows a non-Fermi liquid behavior with a large absolute value and nonlinear temperature dependence both strongly affected by chemical formula (for review of the experimental data see Kaiser and Uher (1991)).

With the canonical Boltzmann equation we find for $2e$ charged bosons

$$S = \frac{k_B}{2e} \left(\frac{\int d\xi \xi \sigma(\xi) sinh^{-2}(\xi/2 - \mu/2T)}{\int d\xi \sigma(\xi) sinh^{-2}(\xi/2 - \mu/2T)} - \frac{\mu}{T} \right) \qquad (7.19)$$

where $\xi = E/T$ is the reduced energy, $\sigma(\xi)$ is the differential conductivity $(\sim E\tau(E))$, which in general depends on energy.

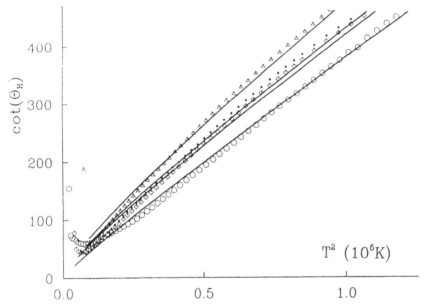

Fig.7.7. $Cot\Theta_H$ compared with the theory.

If the chemical potential is pinned to the mobility edge $\mu \simeq 0$ (see Section 7.2.2) the thermoelectric power determined with Eq.(7.19) is temperature independent. It can be very large $(k_B/2e \simeq 43\mu V/K)$ if $\sigma(\xi)$ has a (super)linear energy dependence, such that the ratio of integrals in Eq.(7.19) is of order unity. Large values of S weakly dependent on temperature are observed in many underdoped copper-based oxides in a wide temperature range above T_c. Optimally doped and overdoped samples can be close to the half filling of the *bipolaron* band $n \simeq 1/2$. Because bipolarons are $hard - core$ bosons one can apply following Mott (1993a) the expression of S for a near half-filled narrow band :

$$S \simeq \frac{k_B}{2e} ln \frac{1-n}{n} \qquad (7.20)$$

which in optimally doped $YBCO$ is of order $5 - 10\mu V/K$. A near half filling can be responsible also for the change of sign of S observed in some copper-based oxides with doping and temperature. If the chemical potential is pinned to the mobility edge localised bipolarons contribute to S with the opposite sign. This contribution also lowers the value of S.

Yu *et al.* (1988) argued that the absence of any effect on S of the magnetic field up to $30T$ is evidence for spinless particles, such as bipolarons.

7.2.5 Infrared conductivity

Infrared absorption and reflection measurements provide an insight into the nature of charge carriers and their interaction with the lattice. Studies of photoinduced carriers in the dielectric 'parent' compounds like La_2CuO_4, $YBa_2Cu_3O_6$ and others demonstrate the formation of self-localised small polarons or bipolarons (Kim *et al.* (1988), Taliani *et al.* (1990)). In these experiments the sample was pumped by a laser beam and the net change in the absorption coefficient was determined from the photoinduced change in transmission. New photoinduced phonon modes were found indicating the formation of a localised structural distortion around a photogenerated carrier. In addition, a broad peak of photoinduced absorption in the electronic region of frequencies was observed, indicating the formation of localised electron states deep inside the semiconducting energy gap. These two aspects of the data confirm the formation of self-localised polarons and provide direct evidence of the importance of the electron-phonon interaction in metal oxides. Mihailovic *et al.* (1990) described the spectral shape of the photoconductivity with the small polaron transport theory. They also argued that the similar spectral shape and systematic trends in both photoconductivity of optically doped dielectric samples and infrared conductivity of chemically doped high$-T_c$ oxides indicate that carriers in the concentrated (metallic) regime retain much of the character of carriers in the dilute (photoexcited) regime implying that the charge carriers in the normal state of high-T_c cuprates are polarons or bipolarons.

Another piece of evidence for bipolarons in high -T_c oxides and their Bose-Einstein condensation comes from the near infrared absorption in the high frequency region ($\nu \sim 0.5 - 0.7eV$) (Alexandrov *et al.* (1993b)). It is well known that for frequencies much higher than the superconducting energy gap no change with temperature in optical absorption is expected within the BCS theory. However Dewing and Salje (1992) observed the effect of the superconducting phase transition on the near infrared absorption with the characteristic frequency $\nu \simeq 0.7eV$ in $YBa_2Cu_3O_{7-\delta}$, Fig.7.8. The relative change of the integrated optical conductivity (frequency window $2000cm^{-1} - 10000cm^{-1}$) at the superconducting transition was as high as 10%.

To explain the effect we derive the sum rule for the bipolaron optical conductivity. Because the Fröhlich interaction is diagonal in the site-representation, Eq.(4.12) it does not change the conductivity sum rule (Kubo (1957)), for a single band Hubbard model derived by Maldague (1977) :

$$I(T) = \frac{\pi e^2 a^2}{2} \langle - \sum_{\mathbf{m},\mathbf{m}',s} T(\mathbf{m} - \mathbf{m}') c^\dagger_{\mathbf{m},s} c_{\mathbf{m}'s} \rangle, \qquad (7.21)$$

where a is a lattice constant and

$$I(T) = \int_0^\infty d\nu \sigma(\nu) \qquad (7.22)$$

is the integrated conductivity. To calculate the kinetic energy of the strongly coupled electron-phonon system up to the second order in the transfer integral with the accuracy $\sim (T(\mathbf{m})/E_p)^2 \sim 1/\lambda^2$ one can apply the polaronic S, Eq.(4.16) and bipolaronic S_2, Eq(5.34) canonical transformations to Eq.(7.21) with the following result:

$$I(T) = \frac{\pi e^2 a^2}{2} \left(2v^{(2)}(n - \langle n_\mathbf{m} n_{\mathbf{m}+\mathbf{a}} \rangle) + 2 \langle \sum_{\mathbf{m},\mathbf{m}'} t_{\mathbf{m},\mathbf{m}'} b^\dagger_\mathbf{m} b_{\mathbf{m}'} \rangle \right), \qquad (7.23)$$

with $v^{(2)}$ determined in Section (5.2) and n the bipolaron atomic density.

The conductivity sum rule, Eq.(7.23), includes optical absorption due to the bipolaron dissociation with and without phonon emission and absorption. It includes also the low-frequency Drude conductivity of small bipolarons tunneling in a bipolaronic narrow band (the half-bandwidth t) due to their scattering by phonons, impurities and by each other without their dissociation.

If one is interested only in the optical part I_{opt} ($\nu \simeq \Delta$ or higher) one should subtract the bipolaronic Drude contribution I_{Drude} from the total absorption intensity:

$$I(T) = I_{opt}(T) + I_{Drude}(T). \qquad (7.24)$$

To derive the integrated bipolaronic Drude conductivity, I_{Drude} one can apply the sum rule to the bipolaronic Hamiltonian, Eq.(5.47), keeping in mind that bipolarons have charge $2e$:

$$I_{Drude}(T) = \frac{4\pi e^2 a^2}{2} \langle \sum_{\mathbf{m,m'}} t_{\mathbf{m,m'}} b_{\mathbf{m}}^{\dagger} b_{\mathbf{m'}} \rangle \qquad (7.25)$$

Subtracting Eq.(7.25) from Eq.(7.23) one obtains:

$$I_{opt} = \pi e^2 a^2 \left(v^{(2)}(n - \langle n_{\mathbf{m}} n_{\mathbf{m+a}} \rangle) - \langle \sum_{\mathbf{m,m'}} t_{\mathbf{m,m'}} b_{\mathbf{m}}^{\dagger} b_{\mathbf{m'}} \rangle \right).$$
$$(7.26)$$

In a homogeneous repulsive Bose-liquid the short range pair correlation function $g(r) = \langle n(0)n(r) \rangle/n^2$ is small. As example in liquid He^4 $g(r) < 0.2$ for $r < 2.5A$. The temperature dependent part of this correlator is even smaller. Thus the contribution to the conductivity sum rule of the term quadratic in the bipolaron density is practically temperature independent and negligible for the homogeneous strongly repulsive Bose-liquid. One can also neglect the corrections to the free particle energy spectrum, which are small while r_s is small. These simplifications yield:

$$I_{opt}(T) = \pi e^2 a^2 \left((v^{(2)} - t)n(T) + \int_0^{2t} \frac{\epsilon N_b(\epsilon) d\epsilon}{y^{-1} exp(\epsilon/T) - 1} \right), \quad (7.27)$$

where $N_b(\epsilon)$ is the density of states in a narrow bipolaronic band, and $n(T)$ is the bipolaron atomic density determined by the chemical potential μ ($y = exp(\mu/T)$).

The first term in Eq.(7.24) describes the incoherent absorption of light accompanied by emission and absorption of many phonons. The frequency dependence of this absorption was discussed by Bryksin and co-workers (1983, 1988). The absorption maximum lies at a frequency $4E_p - V_c$. Comparing the maximum position with experiment, Fig.7.8 (insert) one obtains the realistic value of the polaronic level shift of the order of a few hundreds meV. Bryksin *et al.* ignored the tunneling motion of bipolarons, so their results are not valid in the temperature range compared with T_c as far as the temperature dependence of the absorption intensity is concerned. But the gross spectral features on the frequency scale of the order of E_p are, of course, temperature independent.

The second term in the integrated absorption, Eq.(7.27) is due to the coherent tunneling contribution. This term is responsible for the influence of the superconducting transition on the optical absorption.

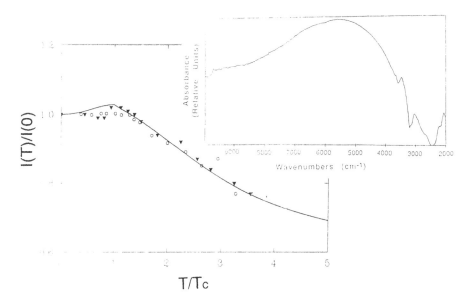

Fig.7.8. Integrated optical absorption of inter-site bipolarons (solid line) and experiment, $YBa_2Cu_3O_7$ (triangles) and $YBaCuO/Fe$ (circles). Insert: the spectral dependence of the near-infrared absorption.

In the limit of a very high characteristic phonon frequency $\omega >> \Delta$, phonons can not be emitted or absorbed in the relevant frequency range. In this limit $v^{(2)} = t$, Eq.(5.42). If $v^{(2)} = t$ the incoherent contribution to I_{opt} turns to be zero, and at zero temperature $I_{opt}(0) = 0$. We conclude that the *condensed pairs on a lattice cannot absorb light coherently*. This is because of the parity and time-reversal symmetry.

The Coulomb repulsion on copper (Hubbard U) is estimated to be of order of a few eV and on oxygen it is of order of 1eV. Thus the formation of on-site bipolarons seems to be unrealistic because the twice polaronic level shift is less than 1eV. However the formation of inter-site small bipolarons on copper and (or) on oxygen is quite feasible. One can generalise the expression for the integrated optical absorption Eq.(7.27) for the case of inter-site bipolarons, assuming that temperature is small for their dissociation, but is comparable with the singlet-triplet exchange energy J. In general the absorption spectra and the integrated optical conductivities in a finite frequency window for singlet and triplet bipolarons are different. That can be due to an additional absorption with the internal transition like the bonding-antibonding transition of a singlet which is absent from a triplet bipolaronic state (see Fig.7.1). While $\Delta \gg T$ the total number of bipolarons is practically temperature independent. With the additional spin quantum number l for bipolaronic singlet ($l = 0$) and triplet ($l = 0, \pm 1$) states $b_{\mathbf{m},l}$ one obtains:

$$\frac{I_{opt}(T)}{I_{opt}(0)} = \delta I_n(T) + \delta I_c(T), \qquad (7.28)$$

where the noncoherent contribution, responsible mainly for the temperature dependence of the normal state absorption is

$$\delta I_n(T) = 1 - \frac{3AT}{2t} ln \left(\frac{1 - y exp(-2t/T - J/T)}{1 - y exp(-J/T)} \right) \qquad (7.29)$$

and the coherent one:

$$\delta I_c(T) = \delta_c \frac{T^2}{T_c^2} \int_0^{2t/T} x dx \left(\frac{1}{y^{-1} exp(x) - 1} + \frac{3}{y^{-1} exp(x + J/T) - 1} \right) \qquad (7.30)$$

with y determined now from the thermal equilibrium of singlet and triplet bipolarons Eq.(7.1). The temperature independent constant A is proportional to the difference of the absorption of the singlet and triplet bipolaron and should be determined from the comparison with experiment. The triplet bandwidth is assumed to be the same as for the singlet and $\delta_c = T_c^2/2nt(v^{(2)} - t)$.

The concentration of singlet intersite bipolarons is strongly temperature dependent on the scale $T \sim J$ because the density of triplet states is high, three times that of singlets. This fact allows us to describe the pronounced decrease of the integrated absorption intensity in the normal state, Fig.7.8 with the reasonable value of the singlet-triplet exchange energy $J/T_c \simeq 4.4$. A remarkable drop of the slope of the optical absorption below T_c, Fig 7.8, is due to the temperature dependent coherent contribution to the sum rule. This contribution, Eq.(7.30) vanishes at $T = 0$ because of the Bose-Einstein condensation of singlet bipolarons. The pronounced influence of the Bose-Einstein condensation on the near infrared absorption shows that bipolarons are responsible for the phenomenon of high T_c.

Within the last years different spectroscopy data have been found consistent with the existence of small (bi)polarons in high-T_c superconductors: the observation by Sugai(1991) of both infrared and Raman-active vibration mode, anomalous midinfrared optical absorption (Thomas *et al.* (1992); Bi and Eklund (1993)), temperature dependence of the vibration energy of atoms engaged in the polaron formation (Mook *et al.* (1990)), XAFS results on the radial distribution function of apex oxygen ions (Mustre de Leon et al (1990)) interpreted as an evidence for intersite pairs (Mustre de Leon *et al.* (1992); Ranninger and Thibblin (1992)). Neutron difraction experiments find that doping introduces dynamic regions of displaced atoms that are consistent with the polaronic carriers (Toby *et al.* (1990)).

7.2.6 C-axis transport

Transport perpendicular to the CuO_2 layers is important for the understanding of the nature of carriers. In optimally doped $YBCO$ c-axis resistivity shows metallic like temperature behavior with the mean-free path comparable or slightly less than the interplane distance. If the interplane hopping integral is very small the normal state transport along the c-axis should be thermally activated. This is usually observed in strongly anisotropic Bi and Tl-based copper oxides at temperature just above T_c because the (bi)polaron band in the c-direction is very narrow in these materials, the bandwidth below a few tens K. In the superconducting state a very small temperature dependent plasma gap was observed with the c-axis optical reflectivity ($\omega_{ps} \simeq 50cm^{-1}$ in $La_{1.84}Sr_{0.16}CuO_4$, Uchida (1992)). The value of the c-axis plasma gap is below the estimated BCS gap 2Δ which is quite anomalous

situation and has never been observed in conventional superconducting materials. The gap disappears above T_c, Fig 7.9, which
plausibly corresponds to a plasma frequency of the condensate.

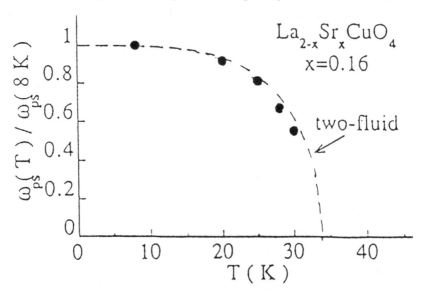

Fig.7.9. Plasma frequency of the condensate plotted as a function of temperature (Ushida (1992)).

We propose a simple interpretation of ω_{ps} based on the Bogoliubov spectrum of a charged condensed Bose-gas (Chapter 1):

$$\epsilon_{\mathbf{k}} = \sqrt{E_{\mathbf{k}}^2 + 2E_{\mathbf{k}}V(\mathbf{k})n_0(T)} \qquad (7.31)$$

where $E_{\mathbf{k}}$ is the anisotropic bipolaron band dispersion, determined
with Eq.(1.40). In the long-wave limit $k_{\parallel} = 0, k_{\perp} \to 0$ $V(\mathbf{k})$
is the three-dimensional Coulomb interaction $V(\mathbf{k}) \sim 1/k^2$ and
the excitation spectrum Eq.(7.31) has a plasma -like temperature
dependent gap

$$\omega_{ps} = \sqrt{\frac{16\pi n_0(T)e^2}{\epsilon_0 m_c^{**}}} \qquad (7.32)$$

with $m_c^{**} = 1/2t_{\perp}d^2$ the effective bipolaron mass in the c-direction.
The gap Eq.(7.32) depends on temperature, disappearing above T_c
because only condensed bosons contribute to it.

7.3   Superconducting state

7.3.1   Isotope effect

The advances in the fabrication of the isotope substituted samples made of possible to measure a sizable isotope effect , $\alpha = -d\ln T_c/d\ln M$ in many high-T_c oxides, including $YBCO$ (Franck *et al.* (1989,1991). This led to a general conclusion that phonons are relevant for high T_c contrary to some earlier suggestions (see e.g. Anderson and Abrahams (1987)). Moreover it is realised now that the isotope effect is very different from the BCS prediction where $\alpha = 0.5$ (or less). Several compounds show $\alpha > 0.5$ (Crawford *et al.* (1990); Franck *et al.* (1991)), and a small negative value of α is found in $Bi - 2223$ (Bornemann *et al.* (1991)). Sometimes it is argued that these striking variations in the isotope effect, including its large values do not necessarily imply a large phonon contribution to the T_c value and do not invalidate an electronic mechanism with a small phonon contribution. However if the two mechanisms (the phonon mediated attraction and the exotic nonphonon one) both make a contribution, then within the BCS approach as the nonphonon interaction dominates with increasing T_c, α should decrease and approach zero at the highest T_c. But in this case α would not go negative as was found in $Bi2223$. With the BCS theory it is hardly possible to obtain a negative value of α if T_c as high as 100K. In all conventional superconductors with anomalously low or negative value of the isotope effect T_c is less than $1K$.

With the formula for T_c derived in Section 4.2 for the polaronic superconductor one can explain all three unusual features of the oxygen isotope effect, in particular large values in low T_c oxides, an overal trend to lower value as T_c increases, and a negative α at high T_c, Fig. 7.10. As far as the oxygen mass dependence is concerned the formula Eq.(5.18) is applied both to the BCS-like polaronic superconductor and to the Bose-Einstein condensation of *intersite* bipolarons with the effective mass depending exponentially on the electron-phonon coupling g: $m^{**} \simeq m^* \sim exp(g^2)$. The only quantity, which depends on the oxygen isotope mass M is g:

$$g^2 = const\sqrt{M} + g_s^2 \qquad (7.33)$$

where g_s^2 is a possible contribution to the mass renormalisation from the vibrations of other ions and from spin fluctuations. Differentiating Eq.(5.18) one obtains

$$\alpha = \frac{\beta g^2}{2} \left(1 - \frac{e^{-g^2}}{\lambda - \mu_c} \right) \tag{7.34}$$

with $\beta = 1 - g_s^2/g^2$ a relative contribution of oxygen to the small polaron cloud. Expressing T_c in units of $\tilde{D}$ we find

$$T_c = exp \left(-g^2 - \frac{e^{-g^2}}{\lambda - \mu_c} \right) \tag{7.35}$$

With the equations (7.34,35) one can analyse the correlation between the critical temperature and the isotope shift, Fig. 7.10.

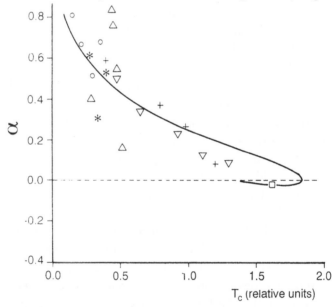

Fig.7.10. Oxygen isotope effect:(o)$La_{2-x}Ca_xCuO_4$, ($\triangle$)$La_{2-x}Sr_xCuO_4$, (*)$La_{2-x}Ba_xCuO_4$ (Crawford *et al.* (1990)); (∇)$Y_{1-x}Pr_xBa_2Cu_3O_{6.92}$ (Franck *et al.* (1991)); (+)$YBa_2Cu_{4-x}Ni_xO_8$ (Bornemann *et al.* (1991a)); ($\square$)$Bi - Pb - Ca - Sr - Cu - O$ (Bornemann *et al.* (1991b)). Theoretical curve after Alexandrov (1992a).

They describe all remarkable features mentioned above. It follows from our considerations that with the isotope effect one can distinguish the BCS like polaronic superconductivity $\alpha < 0$ from the Bose-Einstein condensation of small bipolarons $\alpha > 0$. With the increasing ion mass in the bipolaronic superconductor the bipolaron mass increases and the Bose-Einstein condensation temperature T_c decreases. On the contrary in polaronic superconductors the increase of the ion mass leads to the band narrowing and to the enhancement of the polaron density of states and therefore of T_c. The value of $\beta \simeq 1/3$ obtained from the fit to the experiment, Fig. 7.10, shows that more than 30% of the cloud around polaron comes from oxygen vibrations, the rest is due to vibrations of other ions and (or) due to spin fluctuations, as proposed by Mott (1990). If the last contribution increases with T_c the isotope effect is more suppressed. Overdoping should remove the outer lattice part of polaron and therefore reduce α.

7.3.2 Heat capacity

Another striking feature in which the superfluid Bose liquid differs significantly from the superfluid Fermi liquid is the specific heat near the transition temperature. Bose liquids (or more precisely He^4) show the characteristic λ like singularity in the specific heat. Superfluid Fermi liquids (He^3 and the BCS superconductors), on the contrary, exhibit a sharp second order transition accompanied by a finite jump in the specific heat, Fig. 1.1.

It has been established beyond doubt (Fisher *et al.* (1988), Loram *et al.* (1988), Inderhees *et al.* (1988), Junod *et al.* (1989), Schnelle *et al.* (1990)) that in high T_c superconductors the anomaly in the specific heat spreads to about $|T - T_c|/T_c \sim 0.1$ or larger, Fig. 1.2. The estimations with the canonical gaussian fluctuation theory yield an unusually small coherence volume, Table 7.2, comparable with the unit cell volume, $\Omega \simeq 167 \mathring{A}^3$ in $YBCO$ (Loram *et al.* (1992)). That means that the overlap of pairs is small (if any). Moreover it was stressed by Salamon et al (1990) that the heat capacity anomaly is logarithmic, and consequently cannot be adequately treated by gaussian corrections to the mean field BCS heat capacity.

On the other hand one can rescale the absolute value of the specific heat and the temperature to compare the experimentally determined specific heat of He^4 with that of high-T_c oxides (Alexandrov and Ranninger (1992a)), Fig. 1.1. The specific heat per

Table 7.2: The coherence volume Ω in $\mathring{A}^3$, the in-plane ξ_{ab} and out- of- plane ξ_c coherence lengths derived from a Ginzburg-Landau analysis of the specific heat (Loram *et al.* (1992)).

Compound	Ω	$\xi_{ab}^2,(\mathring{A}^2)$	$\xi_c, (\mathring{A})$
$YBa_2Cu_3O_7$	400	125	3.2
$YBa_2Cu_3O_{7-0.025}$	309	119	2.6
$YBa_2Cu_3O_{7-0.05}$	250	119	2.1
$YBa_2Cu_3O_{7-0.1}$	143	119	1.2
$Ca_{0.8}Y_{0.2}Sr_2Tl_{0.5}Pb_{0.5}Cu_2O_7$	84	70	1.2
$Tl_{1.8}Ba_2Ca_{2.2}Cu_3O_{10}$		40	< 0.9

boson in the two high T_c oxides practically coincides with that of He^4 ($n_b = 1$) in the entire region of the λ singularity. In fact for the 2223 compound the λ shape is experimentally better verified than in He^4 itself because of the fifty times larger value of the critical temperature. The density of *nonlocalised* bosons $n_b(T_c)$ determined from the heat capacity fit to He^4 is very close to that determined from the Hall measurements, Fig.7.5 :

$$n_b \simeq 1.8 \times 10^{21} cm^{-3} \tag{7.36}$$

in the optimally doped $YBCO$. The specific heat of a single crystal of $YBa_2Cu_3O_7$ was measured with the magnetic fields up to $8T$ providing strong evidence for the critical exponents consistent with those observed in He^4 (Overend *et al.* (1994)).

In general it is difficult to analalyse the specific heat above T_c because T_c is so high that the normal state temperature regime is dominated by phonons. However Loram and co-workers (1993) using the differential technique extracted the electronic contribution in the normal state above T_c up to room temperature and found an electronic entropy rather large and linear in temperature in optimally doped $YBCO$. In the underdoped samples they measured smaller entropy and reported unexpected deviations from linearity plausibly connected with the spin gap, Fig. 7.11 (insert). One can calculate the entropy of the extended bosons above T_c with the classical expression for the Bose gas

$$\frac{S_{ext}}{k_B} = \frac{T}{2t} \int_0^1 dx \left(\frac{yln(1/xy - 1)}{1 - xy} - \frac{ln(1 - xy)}{x(1 - xy)} \right) \tag{7.37}$$

where the density of states in the bipolaron band is assumed to be energy independent. Because the chemical potential is pinned to the mobility edge ($y \simeq 1$) as discussed in Section 7.2.3 this expression predicts a practically linear entropy with the absolute value fitting well the experimental observation, Fig. 7.11. The localised bosons contribute also to the electronic entropy; however their contribution depends on the Coulomb repulsion and can be suppressed. The lowering in the value of the entropy above T_c with the increase in oxygen deficiency (Fig.7.11) is due to the localisation by a random field, which is also responsible for the drop of T_c and of the heat capacity jump in Zn doped samples. The deviations of S from linearity in the underdoped samples are explained within our model by the existence of thermally excited triplet bipolarons, which are responsible for the spin gap effect in NMR, neutron scattering and resistivity (sections 7.2.1 and 7.2.3).

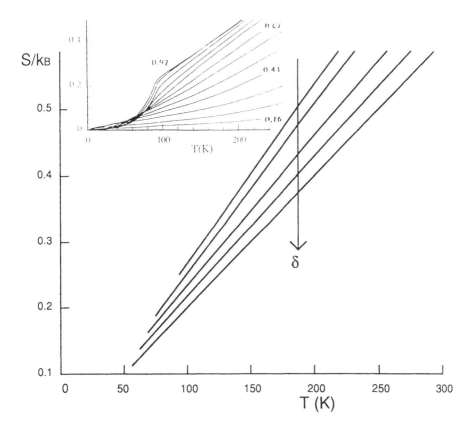

Fig.7.11. Theoretical entropy of extended bosons ($2t = 1000K$) and carrier entropy in $YBa_2Cu_3O_{7-\delta}$ (Loram *et al.* (1993), insert); different curves correspond to different values of $x = 1 - \delta$.

7.3.3   Thermal conductivity

Among other properties thermal conductivity shows an unusual temperature behaviour, which is just opposite to that predicted by the standard BCS theory. The wellknown in-plane thermal conductivity enhancement in the superconducting state, which is a general property of high-T_c copper oxides, is commonly attributed to the lattice contribution, which is limited by the phonon-electron scattering. This contribution explains the thermal conductivity enhancement in some low-T_c alloys (Shoenberg (1952)). However in view of the measurements on high-quality crystals (Cohn *et al.* (1992), Yu *et al.* (1992)) this explanation is questioned. Unlike previous analyses, Yu *et al.* attribute the observed rapid rise in the superconducting-state thermal conductivity to the electronic contribution with a strongly suppressed scattering rate. The normal-state thermal conductivity poses also a problem. With the electronic Lorentz number and the experimental value of resistivity one obtains a sizable electronic contribution to the normal-state thermal conductivity, approximately half of the measured value. This clearly contradicts the near equality of the in-plane thermal conductivity above $100K$ in the insulating and $90 - K$ crystals, suggesting a more radical viewpoint that the electronic contribution is negligible in both systems (Hagen *et al.* (1989)). This would be the case according to the Wiedemann- Franz ratio if the carriers have charge 2e. We derive in this section transport relaxation rates for near two- dimensional charged bosons in the normal and superconducting states to explaine the main puzzling features of the normal and superconducting state thermal conductivity of high-Tc oxides (Alexandrov and Mott (1993a)). Our explanation is quite compatible with the phenomenological model proposed by Yu *et al.*. The fundamental difference is that the carriers in our model are bosons with charge 2e. We find an infinite thermal conductivity of near $2d$ bosons below T_c.

The excitation spectrum of a Bose gas is described by the Bogoliubov expression. For two-dimensional 2e charged bosons it is given by Eq.(1.50) with

$$q_s = q_d \left(\frac{n_0(T)}{n} \right)^{1/3}$$

and $q_d = (32\pi e^2 nm^{**}/\epsilon_0)^{1/3}$. The Bogoliubov approximation is valid for a restricted temperature range close to $T = 0$. However, one can extend the mean-field expression Eq.(1.50) up to $T = T_c$ in a qualitative analysis taking into account the depletion of the superfluid component with temperature. For weakly interacting near $2d$ bosons with logarithmic accuracy (see Eq.(1.45))

$$T_c = \frac{2\pi n}{m^{**}L} \tag{7.38}$$

The condensate density in a wide temperature range not very close to zero is given by the free boson expression

$$n_0(T) = n(1 - t) \tag{7.39}$$

Here $t = T/T_c$ is a reduced temperature and n the in-plane density.

The elastic scattering of excitations is described by the Hamiltonian:

$$H_s = \sum_{\mathbf{k},\mathbf{k'}} v(\mathbf{k}, \mathbf{k'}) \alpha_{\mathbf{k'}}^{\dagger} \alpha_{\mathbf{k}} \tag{7.40}$$

with

$$v(\mathbf{k}, \mathbf{k'}) = \frac{v_0(\mathbf{k} - \mathbf{k'})(u_{\mathbf{k}} u_{\mathbf{k'}} + v_{\mathbf{k}} v_{\mathbf{k'}})}{\epsilon(\mathbf{k} - \mathbf{k'}, 0)} \tag{7.41}$$

a screened scattering potential and the coherence factors $u_{\mathbf{k}}, v_{\mathbf{k}}$ determined from Eq.(1.18-20). Here $v_0(\mathbf{q})$ is a Fourier component of a bare (unscreened) single boson -impurity or boson- acoustic phonon interaction . The latter can be treated as practically elastic if the temperature is not extremely low: $T > T^*$, where $T^* = m^{**}s^2/2$ with s being the sound velocity. T^* is less than 10K for relatively heavy bosons with $m^{**} = 10 m_e$. The static dielectric function of charged bosons $\epsilon(q, 0)$ depends on q below T_c in a rather complicated way. The long wave behavior is approximated by

$$\epsilon(q, 0) = 1 + const \left(\frac{q_s}{q}\right)^{\beta} \tag{7.42}$$

For free $2d$ bosons $\beta = 3, const = 1$ (Hines and Frankel (1979)). In the normal state one can take $\epsilon(q, 0) \simeq 1$ because the relevant momentum transfer q is of order of the boson momentum itself, which is large compared with q_d for $T > T_c$ while r_s is small.

With the Fermi golden rule and the Boltzmann equation one obtains the elastic transport relaxation rate for excitations in the usual way:

$$1/\tau(k) = 2\pi \sum_{\mathbf{k'}} \frac{k_x - k_x'}{k_x} v^2(\mathbf{k}, \mathbf{k'}) \delta(\epsilon_\mathbf{k} - \epsilon_\mathbf{k'}) \tag{7.43}$$

Substitution of Eq.(7.41) into Eq.(7.43) yields:

$$1/\tau(k) = \frac{k\, dk}{\pi\, d\epsilon_\mathbf{k}} \left(u_\mathbf{k}^2 + v_\mathbf{k}^2\right)^2 \int_0^\pi d\phi(1-\cos(\phi)) \frac{v_0^2(k\sqrt{2(1-\cos(\phi))})}{\epsilon^2(k\sqrt{2(1-\cos(\phi))}, 0)} \tag{7.44}$$

First we calculate the integral, Eq.(7.44) in the normal state or for the high-energy excitations with $k > q_s$ in the superconducting state. In these cases $\epsilon(q, 0) = 1$, $\epsilon_\mathbf{k} = k^2/2m^{**}$, and $u_\mathbf{k} = 1, v_\mathbf{k} = 0$. For the scattering by acoustic phonons (or by point defects) v_0 is independent of q ($v_0^2 = C_{ac}t$) and τ_{ac} is independent of k:

$$\tau_{ac} = \frac{1}{m^{**}C_{ac}t}, \tag{7.45}$$

where C_{ac} is a temperature independent constant. For charged impurities $v_0^2(q) = C_{im}/q^2$ and the relaxation time shows a canonical Coulomb scattering behaviour, increasing with energy:

$$\tau_{im}(k) = \frac{2k^2}{m^{**}C_{im}} \tag{7.46}$$

with C_{im} being proportional to the number of impurities.

In the superconducting state for low-energy Bogoliubov excitations with $k < q_s$ and $\epsilon(k) = E_s\sqrt{k/q_s}$ we obtain:

$$\tau_{ac}^s(k) \sim \left(\frac{q_s}{k}\right)^{2\beta - 3/2} \tag{7.47}$$

and

$$\tau_{im}^s(k) \sim \left(\frac{q_s}{k}\right)^{2\beta - 7/2} \tag{7.48}$$

If $\beta > 3/4$ both τ_{ac} and τ_{im} show an enhancement in the superconducting state for low energy excitations, compared with the normal state. This enhancement is explained by a large group velocity of the $2d$ Bogoliubov mode, $d\epsilon/dk$, which is divergent as $k^{-1/2}$ in the long-wave limit and by the screening

To calculate the thermal conductivity K_n due to carriers in the normal state one can apply the standard kinetic theory, developed for metals and semiconductors, replacing the Fermi distribution function by the Bose function. As a result one obtains the Wiedemann-Franz law:

$$K_n = L_B \sigma T \tag{7.49}$$

where $\sigma = 4e^2 n \tau_{ac}/m$ is the normal state conductivity,

$$L_B = (\frac{k_B}{2e})^2 \frac{3B_0(z)B_2(z) - 4B_1^2(z)}{B_0^2(z)} \tag{7.50}$$

is a bosonic Lorentz number, and

$$B_\nu(z) = \int_0^\infty \frac{x^\nu dx}{exp(x - z) - 1} \tag{7.51}$$

with zT the chemical potential. The scattering by acoustic phonons is assumed. In the classical high-temperature limit, $T >> T_c$ we obtain:

$$L_B = 2(\frac{k_B}{2e})^2 \tag{7.52}$$

The same numerical coefficient $(= 2)$ is obtained for $3d$ nondegenerate carriers , scattered by acoustical phonons. The boson Lorentz number, Eq.(7.52), should be compared with the electron one, $L_e = \pi^2 k_B^2/3e^2$ which does not depend on the scattering mechanism and on the dimensionality for degenerate carriers. Their ratio is very small mainly due to the double elementary charge of a boson and is given by

$$\frac{L_B}{L_e} = \frac{6}{4\pi^2} \tag{7.53}$$

which is approximately 0.152.

Taking this into account one can explain the near equality of the thermal conductivity of superconducting and insulating crystals of $YBCO$ in the temperature range above 100K. With the electronic Lorentz number one obtains the carrier temperature independent contribution of approximately 4W/mK (a-direction), which is nearly half of the total one, Fig. 7.12 (insert). On the other hand with the bozonic Lorentz number the carrier contribution is negligible. The mean free path of phonons in both crystals might be of the same value, because the thermal phonons practically are not scattered by bosons. The number of bosons which can scatter thermal phonons is exponentially small, being proportional to $exp(-T/T^*)$. We note that the near equality of thermal conductivity of superconducting and insulating crystalls above 100K also eliminates the possibility of *holons* with charge e.

With lowering temperature the bosonic Lorentz number falls. At $T = T_c$, the chemical potential $z = 0$ and $B_0(z) = \infty$ in Eq.(7.50). If one takes into account three dimensional corrections, then $B_0(0) = L$, and L_B is logarithmically small at $T = T_c$.

The situation changes drastically in the superconducting state. Due to the above mentioned singularity of the group velocity, which is a common feature of surface waves, the $2d$ Bogoliubov mode is a perfect heat carrier. In fact, the thermal conductivity is infinite, if the exponent β of the dielectric constant is larger than $5/4$ for the short-range scattering, or $9/4$ for the Coulomb one. To show this we write the expression for the heat flow, taking into account that in the superconducting state both the chemical and the electrical potentials are zero:

$$\mathbf{Q} = -\sum_{\mathbf{k}} \frac{d\epsilon_{\mathbf{k}}}{d\mathbf{k}} \epsilon_{\mathbf{k}} \frac{\partial f_{\mathbf{k}}}{\partial T} \tau^s(k) \left(\frac{d\epsilon_{\mathbf{k}}}{d\mathbf{k}} \nabla T \right) \qquad (7.54)$$

where $f_{\mathbf{k}}$ is the Bose-Einstein function with zero chemical potential and

$$\tau^s = \frac{\tau_{ac}^s \tau_{im}^s}{\tau_{ac}^s + \tau_{im}^s} \qquad (7.55)$$

Substitution of Eq.(7.55) into Eq.(7.54) yields the superconducting state thermal conductivity:

$$K_s = K_{s0} \frac{(1-t)^{8\beta/3-2}}{t^{4\beta-4}} \int_0^\infty \frac{dx\, x^{10-4\beta}}{sinh^2(x)(x^4 + \eta(1-t)^2/t^5)} \qquad (7.56)$$

where K_{s0}, η are temperature independent constants, $\eta \sim C_{im}/C_{ac}$ depends on the number of impurities and therefore on the quality of the crystal. We replaced the upper limit in the integral of Eq.(7.56), which is E_s/T , by the infinity, because the integration region is restricted by the distribution function and because of the power law singularity of the integrated function if $\beta > 5/4$ and small η. For $\beta > 9/4$ K_s is infinite. This is quite unexpected compared with the usual s-wave BCS superconductor, which has exponentially suppressed thermal conductivity due to a gap in the excitation spectrum.

Three-dimensional corrections to the spectrum restrict the value of K_s. One can also obtain the finite value of the thermal conductivity with the lower value of β in the screening. As example for $\beta = 3/2$ K_s is finite for any finite value of η. The low temperature behaviour of K_s in this example is given by $(t << 1)$

$$\frac{K_s}{K_{s0}} = \frac{96\zeta(5)t^3}{\eta} = \frac{100t^3}{\eta} \tag{7.57}$$

Close to T_c we obtain from Eq.(7.56) $(1 - t << 1)$

$$\frac{K_s}{K_{s0}} = \frac{(1-t)^{3/2}}{\eta^{1/4}} \tag{7.58}$$

In the intermediate temperature region K_s has a maximum, Fig. 7.12, which depends on the impurity concentration, but the position of the maximum $(T \simeq 0.4T_c)$ remains practically unchanged.

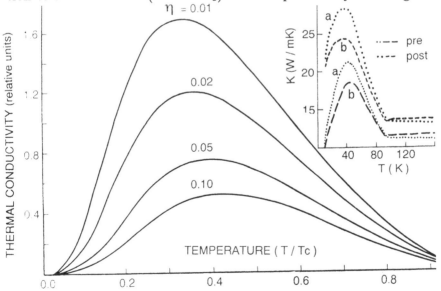

Fig.7.12. Thermal conductivity of near 2d charged bosons. Insert: In-plane thermal conductivity of $YBa_2Cu_3O_{7-\delta}$ (Cohn *et al.* (1992)).

The shape of the thermal conductivity curve of near $2d$ bosons below their condensation temperature, Fig.7.12 is controlled by the simultaneous increase of 'diffusivity' (due to the screening of the scattering potential and the long-wave singularity of the group velocity $\frac{d\epsilon}{dk} \sim k^{-1/2}$) and decrease of the heat capacity $C \sim T^4$. These findings are in global qualitative agreement with the experimental data, Fig.7.12 (insert). For a quantitative comparison one has to include three dimensional corrections and subtract the lattice contribution, which is less than thirty percent of the total value in the superconducting region, as one can estimate using the measurements by Hagen *et al.* (1989) on insulating crystals of $YBCO$. This contribution is a smooth function below T_c. Thus it does not influence the main unusual features of the superconducting heat transport: a sharp rise just below T_c with a maximum approximately at half T_c, and a power-law fall in the low-temperature region.

The enhancement effect of the thermal conductivity in a quasi two- dimensional bipolaronic crystal discussed above is not influenced by the approximation made. As an example bosons must have a hard core. However the enhancement effect is due to long-wave excitations, which are not influenced by a hard core.

7.3.4 London penetration depth

A weak magnetic field penetrates into a charged Bose-gas to a depth (Chapter 1)

$$\frac{1}{\lambda_H^2(T)} = \frac{4\pi e^{*2} n_0(T)}{m^{**}} \qquad (7.59)$$

where for bipolarons $e^* = 2e$. In a near $2d$ Coulomb Bose-gas with $r_s < 1$ the condensate density is linear in temperature $n_0 \sim (1 - T/T_c)$ and so $1/\lambda_H^2$. At low temperature three dimensional corrections to the Bogoliubov spectrum becomes important and can be responsible for the deviation of $\lambda_H^{-2}(T)$ from linearity. The general consensus is that the London penetration depth in copper based high-T_c oxides does not follow the BCS formula and in a wide temperature range has a power law behavior $\lambda_H^{-2} \sim 1 - (T/T_c)^\nu$ with the exponent ν close to unity in $YBCO$ in the intermediate temperature range. Extrapolating the Hall constant R_H above

T_c, Fig.7.5 to zero temperature one determines the density of extended bosons at $T = 0$, which is close to the condensate density for weakly interacting bosons. With $n_0 = n_H(0) = 1/R_H 2e$ and with the experimentally determined low temperature $(T \simeq 4K)$ values of the in-plane and out-of- plane $\lambda_H(0)$ one obtains the low-frequency mass enhancement both in the ab and c directions, Fig.7.13. The value of the in -plane effective mass depends on the composition and increases with the oxygen depletion being in all samples of $YBCO$ larger than $10m_e$. The out-of plane mass is more than one order of magnitude larger. These observations are in line with the polaronic mass enhancement.

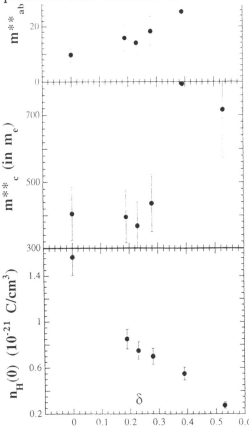

Fig.7.13. In-plane and out- of- plane effective masses of the bipolaron as measured from the London penetration depth and the density of extended bosons $(T = 0)$ determined from the Hall effect (Athanassopoulou N and Cooper J R 1994, unpublished).

Uemura *et al.* (1991) observed with μsR that T_c of many high-

T_c materials scales linearly with $\lambda_H^{-2}(0)$ which should be the case if carriers are near $2d$ bosons preformed above T_c according to Eq.(1.45).

7.3.5   Upper critical field

Possibly the most striking evidence of the charged Bose-liquid in high-T_c oxides comes from the unusual temperature dependence of the upper critical field H_{c2}, which was predicted by one of us (Alexandrov (1984)).

An ideal charged Bose-gas in a magnetic field cannot be condensed because of the one-dimensional character of particle motion within the lowest Landau level (Schafroth (1955)). However the interacting charged Bose-gas is condensed in a field lower than a certain critical value H^* because the interaction with impurities or between bosons broadens the Landau levels and thereby eliminates the one-dimensional singularity of the density of states. The critical field of Bose-Einstein condensation has an unusual positive curvature near T_{c0}, $H^*(T) \sim (T_{c0} - T)^{3/2}$ and diverges at $T \to 0$, where $T_{c0} \simeq 3.3n^{2/3}/m$ is the critical temperature of Bose-Einstein condensation of an ideal gas in zero field.

H^* is determined as the field in which the first nonzero solution of the linearized stationary equation (5.104) for the macroscopic condensate wave function $\psi_0(\mathbf{r})$ appears:

$$\left(-\frac{1}{2m^{**}}(\nabla - 2ie\mathbf{A}(\mathbf{r}))^2 + U_{imp}(\mathbf{r})\right)\psi_0(\mathbf{r}) = \mu\psi_0(\mathbf{r}), \quad (7.60)$$

where $2e$ is the charge of a boson, $\mathbf{A}(\mathbf{r})$, $U_{imp}(\mathbf{r})$ and μ are the vector, random, and chemical potentials, respectively.

The definition of H^*, Eq.(7.60) is identical to that of the upper critical field H_{c2} of BCS superconductors of the second kind. Therefore H^* determines the upper critical field of bipolaronic or any 'bosonic' superconductor. We consider here the impurity scattering as the main source of the broadening.

In general the energy spectrum of the hamiltonian, Eq.(7.60) contains discrete levels (localized states) and a continuous part (extended states). The density of extended states $\tilde{N}(\epsilon, H^*) \sim Im\Sigma(\epsilon)$ and the lowest extended energy level E_c (the mobility edge : $\tilde{N}(E_c, H^*) = 0$) can be found with the random phase ('ladder') approximation for the one-particle self- energy :

$$\Sigma(\epsilon) = \frac{4\pi^2 n_{im} f^2}{m} \int \frac{N(\epsilon', H^*)d\epsilon'}{\epsilon - \epsilon' - \Sigma(\epsilon)} \tag{7.61}$$

where n_{im} is the impurity concentration, f is the scattering amplitude in zero field, and

$$N(\epsilon, H) = \frac{\sqrt{2}(m^{**})^{3/2}\omega}{4\pi^2} Re \sum_{N=0}^{\infty} \frac{1}{\sqrt{\epsilon - \omega(N + 1/2)}} \tag{7.62}$$

is the density of states for a noninteracting system with $\omega = \frac{2eH^*}{m^{**}}$.
  The solution of Eq.(7.61) yields:

$$\begin{aligned}
\tilde{N}_0(\epsilon, H^*) &= \frac{\sqrt{6}(m^{**})^{3/2}\omega}{8\pi^2\sqrt{\Gamma_0}}([\frac{\tilde{\epsilon}^3}{27} + \frac{1}{2} + \sqrt{\frac{\tilde{\epsilon}^3}{27} + \frac{1}{4}}]^{1/3} \\
&- [\frac{\tilde{\epsilon}^3}{27} + \frac{1}{2} - \sqrt{\frac{\tilde{\epsilon}^3}{27} + \frac{1}{4}}]^{1/3}),
\end{aligned} \tag{7.63}$$

and

$$E_c = \frac{\omega}{2} - \frac{3\Gamma_0}{2^{2/3}} \tag{7.64}$$

with $\Gamma_0 = (n_{im}8\pi f^2 e H^*)^{2/3}/2m^{**}$ and $\tilde{\epsilon} = (\epsilon - \omega/2)/\Gamma_0$.
  Eq.(7.63) describes the energy dependence of the density of states of the lowest Landau level $(N = 0)$ near the mobility edge. Since the square root singularity of the density of states of upper levels is integrated out (see below) one can neglect their quantization using the zero field density of states for $\epsilon > \omega$:

$$N(\epsilon) \simeq \frac{(m^{**})^{3/2}\sqrt{\epsilon}}{\sqrt{2}\pi^2}. \tag{7.65}$$

  The first nontrivial *extended* solution of Eq.(7.60) appears at $\mu = E_c$. Thus $H^*(T)$ is determined from the conservation of the number of particles n under the condition that the chemical potential coincides with the mobility edge:

$$\int_{E_c}^{\infty} \frac{\tilde{N}_0(\epsilon, H^*)d\epsilon}{exp(\frac{\epsilon - E_c}{T}) - 1} = n\left(1 - (T/T_{c0})^{3/2} - \frac{n_L(T)}{n}\right), \tag{7.66}$$

where $n_L(T)$ is the number of localized bosons.

The left hand side of Eq.(7.66) is the number of bosons on the lowest Landau level, while the second term of the right hand side is the number of bosons on all upper Landau levels, calculated with the classical density of states, Eq.(7.65).

Substitution of Eq.(7.63) into Eq.(7.66) yields the final expression for the critical field of the Bose-Einstein condensation:

$$H^*(T) = H_d(T_{c0}/T)^{3/2} \left(1 - (T/T_{c0})^{3/2} - \frac{Tn_L}{\gamma n}\beta(T/\gamma)\right)^{3/2},$$
(7.67)

with

$$\beta(x) = \sum_{k=0}^{\infty} \frac{(-1)^k}{x+k},$$
(7.68)

and temperature independent $H_d = \phi_0/2\pi\xi_0^2$. The 'coherence' length ξ_0 is determined by both the mean free path $l = (4\pi n_{im} f^2)^{-1}$ and the inter-particle distance:

$$\xi_0 \simeq 0.8(l/n)^{1/4}.$$
(7.69)

$\phi_0 = \pi/e$ is the flux quantum.

The localization does not change the positive '3/2' curvature of the critical magnetic field near T_c. We believe that this curvature is a universal feature of a charged Bose-gas, which does not depend on a particular scattering mechanism and on approximations made. The number of bosons at the lowest Landau level is proportional to the density of states near the mobility edge $\tilde{N}_0 \sim H/\sqrt{\Gamma(H)}$, where the 'width' of the Landau level is also proportional to the same density of states $\Gamma(H) \sim H/\sqrt{\Gamma(H)}$. Hence $\Gamma(H) \sim H^{2/3}$ and the number of condensed bosons is proportional to $H^{2/3}$. On the other hand this number in the vicinity of T_c should be proportional to $T_c - T$ (the total number minus the number of thermally excited bosons). That gives the '3/2' law for H^*.

At low temperature $T \ll \gamma$ the temperature dependence of H^* turns out to be drastically different for different impurity concentration, Fig.7.14 (Alexandrov (1993)).

An upward curvature of H_{c2} near T_c has been observed in practically all superconducting oxides, including cubic ones. The transition in a magnetic field for a wide temperature range starting from mK-level up to T_c has been reported by Mackenzie *et al.* (1993). Resistively determined H_{c2} values from $T/T_c = 0.0025$ to $T/T_c = 1$ in a $T_c = 20K$ single crystal of $Tl_2Ba_2CuO_6$ follow a temperature dependence that is in good qualitative agreement with the type of curve shown in Fig.7.14 for $n_L/n \simeq 1$, Fig. 7.15. Osofsky *et al.* (1993, 1994) also observed the divergent upward temperature dependence of the upper critical field $H_{c2}(T)$ for thin $BSCO$ films, which was 5 times that expected for a conventional superconductor at the lowest temperature. To describe the data they applied the formula Eq.(7.67), Fig. 7.16.

The $(T_c - T)^{3/2}$ dependence of H_{c2} near T_c is remarkably different from that predicted with the canonical Ginzburg-Landau theory $H_{c2} \sim T_c - T$ while the linearised equation (7.60) for the order parameter looks the same. The difference comes from the temperature dependence of the chemical potential μ, which in the case of bosons is determined through the number of bosons, Eq.(7.66) while in the Ginsburg-Landau theory the corresponding coefficient is fixed $(\sim T - T_c)$. This analogy allows us to apply the Ginsburg-Landau functional also to the charged Bose-gas but with an unusual temperature dependence of the coefficients in the expansion of the free energy.

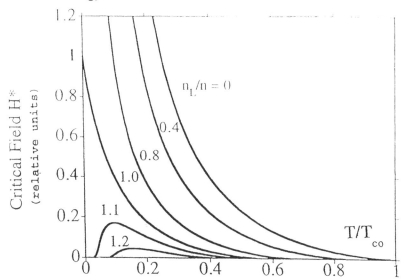

Fig.7.14. Temperature dependence of the critical magnetic field of the Bose-Einstein condensation in units of $H_d(T_{c0}ln2/\gamma)^{3/2}$ for different relative number of localized states n_L/n and $\gamma/T_{c0} = 0.2$.

A charged Bose-liquid of bipolarons turns out to be a simple but a far-reaching model of $H_{c2}(T)$ of both $BiSrCuO$ and $TlBaCuO$.

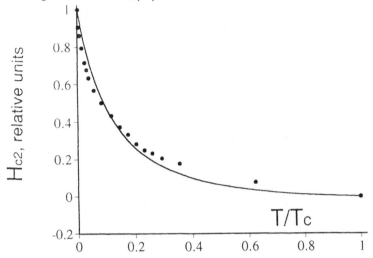

Fig.7.15. The upper critical field of $Tl_2Ba_2CuO_6$ compared with the theoretical curve for $n_L = n$ and $\gamma/T_{c0} = 0.2$.

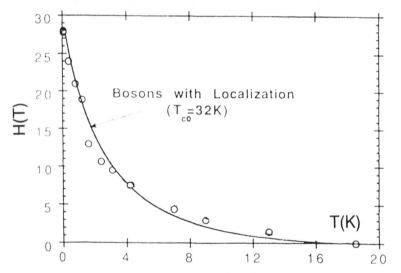

Fig.7.16. The upper critical field of $Bi_2Sr_2CuO_y$ compared with the theory (Osofsky et al (1993)).

The fact that in some oxides the heat capacity ratio γ is constant above T_c should not be used against the model, because the entropy and the specific heat of near two dimensional bosons is perfectly linear above T_c, section 7.3.2. The Van Vleck paramagnetism of intersite singlet bipolarons can contribute to the magnetic susceptibility at low enough temperatures where the spin susceptibility of triplets is low.

7.3.6 Gap, superconducting order parameter and coherence length

There are two gaps for charge excitations in the bipolaronic superconductor. The largest one is the binding energy of a single bipolaron of order $0.1eV$. This gap is temperature independent on the scale compared with T_c, exists *above* T_c and its value does not depend directly on the value of T_c (Alexandrov and Ray (1991)). The large $(2\Delta/T_c \simeq 7 - 8$ for $T_c \simeq 90K)$ temperature and oxygen independent gap existing *above* T_c was found by Schlesinger *et al.* (1991) and by Demuth *et al.* (1990) in $YBCO$ using the infrared reflectivity and the electron energy loss spectroscopy. . The earlier tunneling experiments by Geerk *et al.* (1989) also revealed a large gap disappearing with temperature by weakening rather than by a shift of the dI/dV peak to lower voltages as predicted by the BCS theory.

The second smaller gap is the plasma-like gap in the Bogoliubov spectrum, Eq.(1.49)

$$\omega_{ps} \sim \sqrt{n_0(T)} \qquad (7.70)$$

This gap is closely related to the superconducting order parameter (n_0), depends on temperature and disappears at $T = T_c$ because the condensate disappears $n_0(T_c) = 0$. Its temperature dependence is influenced by the dimensionality of bosons. In an anisotropic crystal it depends also on the wave vector $\mathbf{k}$ according to the formula (7.31) and can be zero along particular directions if the anisotropy is large. This anisotropy is important for the temperature dependence of the London penetration depth. In the bipolaronic superconductor the symmetry of the collective gap $\omega_{ps}(\mathbf{k})$ should be distinguished from that of the 'internal' wave function of a single intersite bipolaron, which is determined by the relative orbital motion of two polarons. The latter is important for Josephson tunneling but hardly affects the Bose-Einstein condensation. Therefore different experiments can measure two different symmetries.

The coherence length ξ_0, measured with a slope of the upper critical field H_{c2} near T_c and with other techniques (for instance with the heat capacity near T_c, Table 7.2) is very small in high-T_c superconductors $(10 - 20\text{Å})$. The small coherence length of order of 10Å (in-plane) pushes any BCS-like theory to the limit of applicability. High-T_c materials exist, as we believe, beyond this limit. However ξ_0 remains larger than the in-plane lattice constant and one can argue that the mean-field BCS approximation remains valid at least qualitatively. However from our consideration of the upper critical field of a charged Bose gas (section 7.3.5) it follows that the coherence length of charged bosons is also large compared with the interparticle distance:

$$\xi_0 \simeq \left(\frac{l}{n}\right)^{1/4} \tag{7.71}$$

where l is a mean free path, restricted by the boson-impurity or boson-boson collisions.

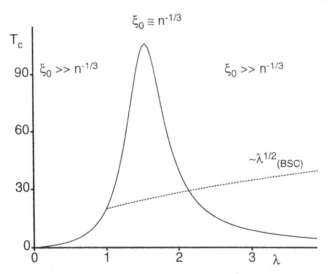

Fig.7.17. T_c (arb. units) and the coherence length of the (bi)polaronic superconductor as a function of λ.

Therefore one can not judge from the value of the coherence length what theory should apply : the BCS or bipolaron one.

Only in the intermediate coupling region $\lambda \sim 1$, where T_c reaches its maximum the coherence length is of order of the interparticle distance, $\xi_0 \simeq n^{-1/3}$, Fig. 7.17.

7.4 Angle-resolved photoemission

It was argued that $ARPES$ shows the well-defined Fermi-surface (see for example Olson *et al.* (1990)), which for $\delta = 0.1$ corresponds well with the plane-related Fermi surfaces calculated from band structure for $YBa_2Cu_3O_{7-\delta}$. Aside from this, however, the spectral shape is quite distinct from that observed in normal metals (Sawatzky (1989)) and the dispersion is much more flat than that from the band structure. The asymmetric lineshape, the clear separation of the coherent and incoherent contributions, signalled by a minimum in the spectrum and the band narrowing are explained by the small polaron theory of ARPES (Alexandrov and Ranninger (1992b)). The intensity of the coherent (i.e. angle- dependent) contribution is strongly reduced due to a factor $exp(-g^2)$ which plays the role of a step 'Z' in the Fermi distribution function, while the broad featureless incoherent background is due to the phonon cloud which constitutes a small polaron. To see these features we calculate the one particle electron Green's function under the condition of the polaron band narrowing

$$g(\mathbf{k},\omega_n) = -\frac{1}{2}\sum_{\mathbf{m}}\int_{-1/T}^{1/T} d\tau\, e^{i\omega_n \tau + i\mathbf{k}\cdot\mathbf{m}}\langle\langle T_\tau c_0(\tau)c_{\mathbf{m}}^\dagger\rangle\rangle \qquad (7.72)$$

For convenience we omit spin. Applying the Lang-Firsov canonical transformation and neglecting the residiual polaron-phonon coupling one obtains

$$g(\mathbf{k},\omega_n) = \frac{T}{N}\sum_{\omega_{n'},\mathbf{m},\mathbf{k'}} \frac{\sigma(\mathbf{m},\omega_{n'}-\omega_n)e^{i(\mathbf{k}-\mathbf{k'})\cdot\mathbf{m}}}{i\omega_{n'}-\xi_{\mathbf{k'}}} \qquad (7.73)$$

with the Fourier component $\sigma(\mathbf{m},\omega_n)$ of the correlation function, determined as

$$\sigma(\mathbf{m},\tau) = exp\left(\frac{1}{2N}\sum_{\mathbf{q}}\gamma^2(\mathbf{q})f_{\mathbf{q}}(\mathbf{m},\tau)\right), \qquad (7.74)$$

with

$$f_{\mathbf{q}}(\mathbf{m},\tau) = [cos(\mathbf{q}\cdot\mathbf{m})cosh(\omega_{\mathbf{q}}|\tau|)-1]coth\frac{\omega_{\mathbf{q}}}{2T}+cos(\mathbf{q}\cdot\mathbf{m})sinh(\omega_{\mathbf{q}}|\tau|).$$

Calculating the Fourier component of Eq.(7.74) and substituting it into Eq.(7.73) one obtains for the case of dispersionless phonons $(\omega_{\mathbf{q}} = \omega_0)$ and a constant $\gamma^2(\mathbf{q}) = 2g^2$

$$\begin{aligned}
g(\mathbf{k},\omega_n) &= \frac{e^{-g^2}}{i\omega_n - \xi_{\mathbf{k}}} + \frac{e^{-g^2}}{N}\sum_{l=1}^{\infty}\frac{g^{2l}}{l!}\\
&\times \sum_{\mathbf{k}'}\left(\frac{n_{\mathbf{k}'}}{i\omega_n - \xi_{\mathbf{k}'} + l\omega_0} + \frac{1 - n_{\mathbf{k}'}}{i\omega_n - \xi_{\mathbf{k}'} - l\omega_0}\right) \quad (7.75)
\end{aligned}$$

In the polaron regime the electron Green's function consists of two different contributions. The first coherent term arises from the polaron band motion. The second $\mathbf{k} - independent$ contribution describes the excitations accompanied by the emission and absorption of phonons. It is this term which is responsible for the asymmetric background of the photoemission spectra. We notice that the spectral density in the second term spreads over a wide frequency range of order of the polaron level shift $2g^2\omega_0$ or more. On the contrary the coherent term shows an angular dependence in a frequency range of order of the polaron bandwidth $2w$.

ARPES measures the imaginary part of the retarded Green's function integrated with a Gaussian instrumental resolution function $F(\epsilon, \epsilon')$ and with the Fermi-Dirac distribution function $n_{\mathbf{k}} = n(\epsilon)$

$$I(\mathbf{k},\epsilon) = -\frac{1}{\pi}\int_{-\infty}^{\infty} d\epsilon' n(\epsilon')F(\epsilon, \epsilon')ImG^R(\mathbf{k}, \epsilon'), \quad (7.76)$$

where $F(\epsilon, \epsilon') = (1/\gamma\sqrt{2\pi})exp(-(\epsilon - \epsilon')^2/2\gamma^2)$.

At present the instrumental resolution $\gamma \simeq 10meV$ is of order of the characteristic relaxation rate of small polarons. G^R is obtained using Eq.(7.75) with the substitution $i\omega_n \rightarrow \epsilon + i0^+$. As a result

$$\begin{aligned}
I(\mathbf{k},\epsilon) &= \frac{n(\xi_{\mathbf{k}})e^{-g^2}}{\gamma\sqrt{2\pi}}exp\left(-\frac{(\epsilon - \xi_{\mathbf{k}})^2}{2\gamma^2}\right) + e^{-g^2}\int d\epsilon' F(\epsilon, \epsilon')\\
&\times \sum_{l}\frac{g^{2l}}{l!}N_p(\epsilon' + l\omega_0)n(\epsilon' + l\omega_0) \quad (7.77)
\end{aligned}$$

with N_p the density of states in the polaron band. The theoretical ARPES calculated with the Gaussian form of N_p is shown in Fig 7.18 compared with the experimental one for $Bi_2Sr_2CaCu_2O_8$.

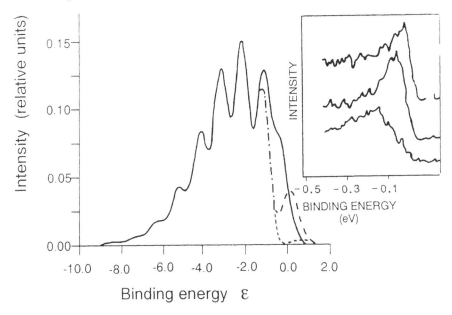

Fig.7.18. Polaron angle-resolved photoemission for three different angles and experiment (insert) in $Bi_2Sr_2CaCu_2O_8$ (Olson *et al.* (1990)); ϵ is measured in units of ω_0.

The broadened asymmetric line shape occurs already at $g^2 = 1$. For $g^2 > 2$ and the bare half bandwidth $D = 200meV$ the polaron bandwidth $2w$ is compared or less than the phonon frequency. In this regime the second term in Eq.(7.77) oscillates as a function of the binding energy. The characteristic *angular* dispersion of ARPES is of order $0.1eV$ which gives an estimate of the polaron bandwidth.

An other puzzling result is a relatively small change in a very large Fermi surface observed when oxygen content was varied in the range $0.5 < \delta < 0.1$, where $YBCO$ is metallic. This is in contrast with the drastic changes in resistivity as indicative of a small Fermi surface, suggesting a more radical viewpoint that the large 'Fermi surface' does not characterize the carriers responsible for the transport in metal oxides (Mott (1993a)). Mott suggested a way round this difficulty, which is illustrated in Fig. 7.19, which shows the one-electron density of states. The shaded area rep-

resents the holes introduced by doping: these form polarons and
eventually bipolarons, so that in fact no states are available in this
area for electrons. In the doped Mott insulator this happens if the
oxygen (hole) band is overlapped with the lowest Hubbard $Cu3d^9$
band.

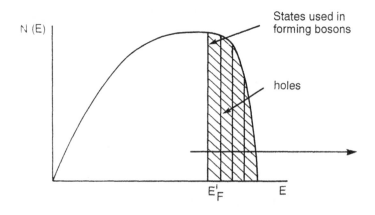

Fig.7.19. Density of states, showing photoemission.

Thus E'_F is not a true Fermi surface, and should not be related
to the conductivity or specific heat. The life-time of a hole left in
the electron background should be proportional to its energy ΔE,
not to $(\Delta E)^2$ as for a true Fermi surface.

7.5   Insulating and overdoped materials

7.5.1   Metal-insulator transition

We complete our discussion of high-T_c oxides with several remarks
on the transition from the non-metallic 'spin glass' phase to the
metallic superconductor, which occurs in for instance $La_{2-x}Sr_xCuO_4$
for $x \simeq 0.12$, and on the overdoped materials, $x > 0.2$. The process is, we believe, similar to that which occurs in doped silicon
for a concentration n_{im} of the dopant given by $n_{im}a_H \simeq 0.26$,

a_H being the radius of orbit describing the trapped carrier (hole). The difference here is that the material before doping is an antiferromagnetic insulator. The first question we have to ask is whether the carriers, trapped small polarons as described earlier in this book, form bipolarons as they do when they become fully mobile. The experimental evidence shows that they do. Analysis by Emin (1992) of the thermopower measured by Fisher *et al.* (1989) shows that the effective charge on a carrier is substantially greater than the electronic charge. Emin finds $2.2|e|$, near enough to $2|e|$. We have then to consider a metal-insulator transition for a gas of trapped bosons. The approach based on the hypothesis of preformed 2e bosons has been used by Doniach and Inui (1990) in a Ginzburg-Landau type analysis of the insulator-superconductor transition in the CuO_2-based materials. Our treatment is based on the assumption, discussed in Section 7.2.2, that charged bosons in a random field will behave like fermions, because their repulsion prevents more than a small number possibly unity, from occupying each potential well. This concept has been used in Section 7.2.3 to explain the Hall effect above T_c.

We next ask whether we should describe the transition to the metallic superconducting state as an Anderson transition, or as a Mott-Hubbard transition depending on a Hubbard U. We note first that, in semiconductor terminology, the system is compensated, because the number of carriers (bosons) is half the number of sites (Sr atoms). Imada (1990) has pointed out that, for a crystalline metal with composition close to that of insulating state, a large mass enhancement is predicted and observed (e.g. V_2O_3 and also $(LaSr)TiO_3$). This is absent in $LSCO$. This suggests that the transition is of Anderson type. In the 'spin glass' state, we think that bipolarons are already formed, but are localized by the disorder produced by the Sr ions. Variable -range hopping is then expected and observed (see for example Tateno (1993)). The discussion given above treats the problem as three dimensional. According to scaling theory in a disordered material in two dimensions all states are localized (Abrahams *et al.* (1979)). Clearly this is not so for the copper oxide superconductors. Another metal-insulator transition of interest is that of $Y_{1-x}Pr_xBa_2Cu_3O_7$; the substitution of Pr for Y, unlike other rare earths, leads to a drop in T_c and eventually to transition to an insulating state. It is thought that the Pr^{+3} produces Anderson localisation. Measurements of resistivity reveal variable range hopping and a positive magnetoresistance, ascribed to a change of the localisation length in a magnetic field (Jiang *et al.* (1994)). In their investigation of the specific heat, Loram *et al.* (1990) show that the substitution of Zn for Cu decreases T_c and the entropy in $YBCO$ up to the transition where T_c eventually drops to zero. Our interpretation is that the additional disorder produced by zinc ensures that all the bosons are Anderson localised.

7.5.2 Overdoped materials

That strong overdoping leads to a drop of T_c to zero, the material remaining metallic, is difficult to explain within the Fermi liquid approach. The concept of overcrouding suggests that the concentration of bosons remains constant, and that of the unbound fermions increases with doping as it was observed by Salje and Güttler (1984) in WO_{3-x}. Then the system is reminiscent of the mixture $He^3 - He^4$ with low density of He^4 atoms and suppressed (or zero) T_c.

Conclusion

The polaronic Fermi-liquid and the bipolaronic Bose-liquid represent a natural extension of the Fermi-liquid and BCS theory to the strong coupling region. In the extreme limit $\lambda \gg 1$ and non-retarded electron-phonon interaction ($\omega > E_F$) the microscopic theory described in this book is reminiscent of the phenomenological model of preformed local pairs, proposed by Ogg (1946) and Schafroth (1955), Butler and Blatt (1955) and discussed in the last decade by many authors using the *negative U* Hubbard hamiltonian. Different from the *negative U* local pair phenomenology the bipolaron theory explains the mechanism of the charged boson formation: bipolarons are formed due to the polaron narrowing of the electron band. This narrowing also explains the high value of T_c. On the other side the well-known shortcomings of the Schafroth model like a huge value of $T_c \simeq 10000K$ are not shared by the bipolaron theory of superconductivity. The enhanced effective mass $m^{**} > 10m_e$ and the low enough concentration $n \simeq 10^{21}cm^{-3}$ push T_c into the range of 100K. The hard-core interaction, which is the only one in the Hubbard U model is irrelevant for the low-energy physics of bipolarons which is controlled by their long-range Coulomb repulsion. This repulsion and the Anderson localisation by disorder are responsible for the unusual kinetic properties of copper-based oxides. The existence of thermally excited triplet bipolarons yields a simple explanation of the spin gap effect in the magnetic susceptibility, resistivity and heat capacity.

The Bose-liquid features of high-T_c oxides lead us to a firm conclusion that a charged Bose-liquid is a simple but far-reaching model of low-frequency kinetics and thermodynamics of these materials. Small polarons and bipolarons are the natural state of carriers in doped semiconductors with a large static dielectric constant and narrow bands. To make them sufficiently mobile high

153

frequency phonons should be involved in the polaron formation. Any bound state is more stable in a low dimensional system. That is why high-T_c (bi)polaronic superconductivity is realised in copper based oxides, which have a large lattice polarisability, high frequency oxygen vibrations and the layer structure.

References

Abrahams E, Anderson P W, Licciardello D C and Ramakrichnan T V 1979 Phys.Rev.Lett. **42**, 673-76.

Abrikosov A A, Gor'kov L P and Dzyaloshinskii I E 1963 Methods of Quantum Field Theory in Statistical Physics, Prentice -Hall, Englewood Cliffs, New Jersey.

Alexandrov A S 1983 Zh.Fiz.Khim. **57**, 273 (Russ.J.Phys.Chem.**57**, 167(1983));
1984 Doctoral thesis, Moscow Engineering Physics Institute (Moscow);
1987 Pis'ma Zh. Eksp.Teor.Fiz.(Prilozh) **46**, 128-31 (1987 JETP Lett.Suppl. **46**, 107-10);
1992a Phys.Rev. B**46**, 14932-35;
1992b Phys.Rev. B**46**, 2838-44;
1992c J.Low Temp. Phys. **87**, 721-29;
1993 Phys.Rev B**48**, 10571-74.

Alexandrov A S, Bratkovsky A M, and Mott N F 1994 Phys.Rev.Lett **72** 1734-37.

Alexandrov A S, Bratkovsky A M, Mott N F and Salje E K H 1993b Physica C**215**, 359-70.

Alexandrov A S and Capellmann H 1991 Phys.Rev. B**43**, 2042-49.

Alexandrov A S, Capellmann H and Göbel U 1992 Phys.Rev. B**46**, 4374.

Alexandrov A S and Elesin V F 1983 Fiz.Tverd.Tela **25**, 456-64 (1983 Sov.Phys.Solid State **25**, 257-62);
1985 Izv.Akad.Nauk,Ser. Fiz. **49**, 326 (1985 Bull.Acad.Sci.USSR

Phys.Ser. **49**(2), 105).

Alexandrov A S, Kabanov V V and Ray D K 1994a Phys.Rev. **B49**, 9915-23.

Alexandrov A S and Kornilovitch P E 1993 Z.Phys. **B91**, 47-50.

Alexandrov A S and Mott N F 1993a Phys.Rev.Lett. **71**, 1075-78; 1993b Invited paper at the MOS-Conference,Eugene, Oregon (USA, July 1993) (1994 J.Supercond.(US) **7**, 599-605).

Alexandrov A S and Ranninger J 1981a Phys.Rev. **B23**, 1796-801; 1981b Phys.Rev. **B24**,1164-69; 1992a Solid St.Commun.**81**, 403-406; 1992b Phys.Rev. **B45**, 13109-12.

Alexandrov A S, Ranninger J and Robaszkiewicz S 1986a Phys.Rev. **B33**, 4526-42; 1986b Phys.Rev.Lett **56**, 949-52.

Alexandrov A S and Ray D K 1991 Phil.Mag.Lett. **63**, 295-302.

Allen P B 1992 Comments Cond.Mat.Phys.**15**, 327-53.

Allen P B and Rainer D 1991 Nature **349**,396-98.

Anderson P W 1975 Phys.Rev.Lett. **34**, 953-56; 1987 Science **235** 1196-98.

Anderson P W and Abrahams E 1987 Nature **327**, 363.

Anderson P W and Schrieffer R 1991 Physics Today **44**, 54-61.

Austin I G and Mott N F 1969 Adv.Phys. **18**, 41-102

Bardeen J, Cooper L N, and Schrieffer J R 1957 Phys.Rev. **108**, 1175-1204.

Bardeen J, Rickayzen G and Tewordt L 1959 Phys.Rev. **113**, 982.

Bassani F, Geddo M, Iadonisi G, and Ninno D 1991 Phys.Rev. **B43**, 5296-306.

Bednorz J G and Muller K A 1986 Z.Phys. **B64**, 189-93.

Blatt J M and Butler S T 1955 Phys.Rev. **100**, 476-80.

Bi X X and Eklund P C 1993 Phys.Rev.Lett. **70**, 2625-28.

Bogoliubov N N 1947 J.Phys.USSR **11**, 23-32.
1958 Zh.Eksp.Teor.Fiz. **34**, 58, 73.

Bornemann H J, Morris D E and Liu H B 1991a Physica C**182**, 132-36.

Bornemann H J, Morris D E, Liu H B, Sinha A P, Narwankar P and Chandrachood M 1991b Physica C**185 − 189**, 1359-60.

Brovman E G and Kagan Yu M 1967 Zh.Eksp.Teor.Fiz. **52**, 557-74 (1967 Sov.Phys.JETP **25**, 365-76).

Bruckel T, Capellmann H, Just W, Scharpf O, Kemmler-Sach S, Kiemel R and Schaefer W 1987 Europhys.Lett. **4** 1189-94.

Bryksin V V and Gol'tsev A V 1988 Fiz.Tverd.Tela **30**,1476-86 (1988 Sov.Phys.Solid.State **30**, 851-56).

Bryksin V V, Voloshin V S, and Raitsev A V 1983 Fiz.Tverd.Tela **25**, 1427-34 (1983 Sov.Phys.Solid.State **25**, 820-24).

Bucher B, Steiner P, Karpinski J, Kaldis E, and Wachter P 1993 Phys.Rev.Lett. **70**, 2012-15.

Carbotte J 1990 Rev.Mod.Phys. **62**, 1027-157.

Carrington A, Walker D J C, Mackenzie A P, and Cooper J R, 1993 Phys.Rev. B**48**, 13051-59.

Chakraverty B K 1981 J.Physique **42**, 1351-56.

Chakraverty B K, Feinberg D, Hang Z, and Avignon M 1987 Solid State Commun. **64**, 1147-51.

Chakraverty B K, Sienko M J and Bonnerot J 1978 Phys.Rev. B**17** 3781-89.

Cohn J L, Skelton E F, Wolf S A, Liu J Z and Shelton R N 1992 Phys.Rev. B**45**, 13144-47.

Cooper L N 1957 Phys.Rev. **104**, 1189.

Crawford M K, Farneth W E, McCarron E M III, Harlow R L and Moudden A H 1990 Science **250**, 1309-18.

Das A N and Sil S 1989 Physica C**161**, 325-30.

De Gennes P G 1989 Superconductivity of Metals and Alloys, Addison-Wesley.

Demuth J E, Persson B N J, Holtzberg F and Chandrashekar C V 1990 Phys.Rev.Lett. **64**, 603-606.

De Raedt H and Lagendijk Ad 1983 Phys.Rev. B**27**, 6097-109.

Dewing H L and Salje E K H 1992 Supercond.Sci.Technol.**5**, 50-53.

Doniach S and Inui M 1990 Phys.Rev. B**41**, 6668-78.

Eagles D M 1966 Phys.Rev. **145**, 645-66.

Eliashberg G M 1960 Zh.Eksp.Teor.Fiz. **38**, 966-76; **39**, 1437-41 (1960 Sov.Phys.JETP **11**, 696-702; **12**, 1000-1002).

Emin D 1970 Phys.Rev.Lett. **25**, 1751-54;
1973 Adv.Phys. **22**, 57; ibid 1975 **24**, 305;
1989 Phys.Rev.Lett. **62**, 1544-47;
1992 in 'Lattice effects in High-T_c Superconductors', eds Bar-Yam Y, Egami T, Mustre-de Leon J and Bishop A R, World Scientific, 377.

Emin D and Holstein T 1969 Ann.Phys.N.Y. **53**, 439-520.

Feinberg D, Ciuchi S and dePasquale F 1990 J.Mod.Phys. **4**, 1317.

Feinman R P, Hellwarth R W, Iddings C K and Platzmann P M 1962 Phys.Rev. **127**, 1004-17.

Fisher D, Genossar J, Parlagen C, Lelong I A and Ashkenazi J 1989 Physica C**162-164**, 1207-08.

Fisher R A, Kim S, Lacy S E, Phillips N E, Morris D E, Markelz A G, Wei J Y T and Ginley D S 1988 Phys.Rev. B**38**, 11942-45.

Fisher M P A, Weichmann P B, Grinstein G and Fisher D S 1989 Phys.Rev. B**40**, 546-70.

Foldy L L 1961 Phys.Rev. **124**, 649-51.

Forro L, Ilakovac V, Cooper J R, Ayache C, and Henry J-Y 1992 Phys.Rev. **B46**, 6626-29.

Franck J P *et al.* 1989 Physica C**162 − 164**, 753-54.

Franck J P *et al.* 1991 Physica C**185 − 189**, 1379-80.

Franck J P, Harker S and Brewer J H 1993 Phys.Rev.Lett. **71**, 283-86.

Friedel J J 1989 J.Phys.:Cond.Mat. **1**, 7757-935.

Friedman L 1964 Phys.Rev. A **135**, 233-46.

Fröhlich H 1950 Phys.Rev. **79**, 845;
1954 Adv.Phys. **3**, 325-61;
1968 Phys.Lett. A**26**, 169-70.

Geerk J, Xi X X and Linker G 1989 Z.Phys. B**73**, 329.

Gehlig R and Salje E 1983 Phil.Mag. **47**, 229-45.

Geilikman B T 1958 Zh.Eksp.Theor.Phys.(USSR) **34**, 1042-44;
1975 Usp.Fiz.Nauk **115**, 403-26 (1975 Sov.Phys.-Usp. **18**, 190-202).

Giaever I 1960 Phys.Rev.Lett. **5**, 147, 464.

Ginzburg V L 1968 Contemp.Phys. **9**, 355-74.

Ginzburg V L and Landau L D 1950 Zh.Eksp.Teor.Fiz. **20**, 1064.

Gogolin A A 1982 Phys.Status Solidi B**109**, 95-108.

Gor'kov L P 1958 Zh.Eksp.Teor. Fiz. **34**, 735-39 (1958 Soviet Phys.JETP **7**, 505-08).

Hagen S J, Wang Z Z and Ong N P 1989 Phys.Rev. B**40**, 9389-92.

Hang Z 1988 Phys.Rev. B**37**, 7419-25.

Hebard A F, Rosseinsky M J, Haddon R C, Murphy D W, Glarum

S H, Palstra T T M, Ramirez A P, Kortan A R 1991 Nature **350**,
  600-01.

Hebel L C and Slichter C P 1959 Phys.Rev. **113**, 1504.

Hertz J A, Fleishman L and Anderson P W 1979 Phys.Rev Lett. **43**, 942-45.

Hines D F and Frankel N 1979 Phys.Rev. **B20**, 972-83.

Hirsch J E 1993 Phys.Rev. **B47**, 5351-58.

Holczer K, Klein O, Huang S M, Kaner R B, Fu K J, Whetten R L, Diederich F 1991 Science **252**, 1154-57.

Holstein T 1959 Ann.Phys. **8**, 325-42; ibid p. 343-89.

Holstein T and Friedman L 1968 Phys.Rev. **165**, 1019-31.

Imada M 1993 J.Phys.Soc.Japan **62**, 1105-108.

Inderhees S E, Salamon M B, Goldenfeld N, Rice J P, Pazol B G and Ginzberg D M 1988 Phys.Rev.Lett. **60**, 1178-81.

Ito T, Takenaka K and Uchida S 1993 Phys.Rev.Lett. **70**, 3995-98.

Jiang W, Peng J L, Hamilton J J and Green R L 1994 Phys.Rev. **B49**, 690-93.

Junod A, Eckert D, Triscone G, Lee V Y and Muller J 1989 Physica **C159**, 215-25.

Kaiser A B and Uher C 1991 in Studies of High Temperature Superconductors, ed. Narlikar A, Nova Science Pub., New York **7**, 353-92.

Kamimura H, Eto M., Matsuno S, Ushio H 1992 Comments Cond. Mat.Phys. **15**, 303-25.

Kim Y H, Foster C M, Heeger A J, Cos S and Stucky G 1988 Phys.Rev. **B38**, 6478-82.

Klein B M, Boyer L L, Papaconstantopoulos D A 1978 Phys.Rev. **B18**, 6411-38.

Klinger M I 1961 Izv.Acad.Nauk SSSR, ser.Fiz. **25**, 342-45.

Klinger M I and Karpov V G 1982 Zh.Eksp.Teor.Fiz. **82**, 1687-703 (1982 Sov.Phys.JETP **55**, 976-84).

Kongeter A and Wagner M 1990 J.Chem.Phys. **92**, 4003-11.

Kubo R 1957 J.Phys.Soc.Japan **12**, 570.

Kubo K and Takada S 1983 J.Phys.Soc.Japan **52**, 2108-17.

Lakkis S, Schlenker C, Chakraverty B K and Buder R 1976 Phys.Rev. **B14**, 1429-40.

Lal R and Joshi S K 1992 Sol.State Commun. **83**, 209-13.

Landau L D 1933 Phys.Z.Sowjetunion **3**, 664.

Lang I G and Firsov Yu A 1962 Zh.Eksp.Teor.Fiz. **43**, 1843-60 (1963 Sov.Phys.JETP **16**, 1301-12).

Little W A 1964 Phys.Rev. **A134**, 1416-24.

Loram J W, Mirza K A and Freeman P A 1990 Physica **C171**, 243.

Loram J W, Mirza K A, Liang W Y and Osborne J 1988 Physica **C162 − 164**, 498-99.

Loram J W, Cooper J R, Wheatley J M, Mirza K A and Liu R S 1992 Phil. Mag. **B65**, 1405-17.

Loram J W, Mirza K A, Cooper J R and Liang W Y 1993 Phys.Rev.Lett. **71**, 1740-43.

Lu Yu, Zhao-Bin Su and Yan-Min Li 1993 Chinese J.Phys. **31**, 579-616.

Ma M, Halperin B I and Lee P A 1986 Phys.Rev. **B34**, 3136-43.

Machi T, Tomeno I, Miyatake T and Tanaka S 1991 Physica **C173**, 32-37.

Mackenzie A P, Julian S R, Lonzarich G G, Carrington A, Hughes S D, Liu R S and Sinclair D C 1993 Phys.Rev.Lett. **71**, 1238-41.

Maldague P F 1977 Phys.Rev. **B16**, 2437-46.

Marsiglio F 1990 Phys.Rev. **B42**, 2416-24;
1993 Phys Lett. **A180**, 280-84.

Massey H S W 1976 'Negative ions', Cambridge University Press.

Matsubara T 1955 Prog.Theor.Phys. **14**, 351-78.

McMillan W J 1968 Phys.Rev. **B167**, 331-44.

Migdal A B 1958 Zh.Eksp.Teor.Fiz. **34**, 1438-46 (1958 Sov.Phys. JETP **7**, 996-1001).

Mihailovic D, Foster C M, Voss K and Heeger A J 1990 Phys.Rev. **B42**, 7989-93.

Mikhailovsky A A, Shulga S V, Karakozov A E, Dolgov O V and Maksimov E G 1991 Soid St.Commun. **80**, 511-15.

Millis A J, Monien H and Pines D 1990 Phys.Rev. **B42**, 167-78.

Mitani N and Kurihara S 1992 Physica C**192**, 230-36.

Monthoux P and Pines D 1992 Phys.Rev.Lett. **69**, 961-64.

Mook H A, Mostoller M, Harvey J A, Hill N W, Chakoumakos B C and Sales B S 1990 Phys.Rev.Lett. **65**, 2712-15.

Mook H A, Yethiraj M, Aeppli G, Mason T E, and Armstrong T 1993 Phys.Rev.Lett. **70**, 3490-93.

Morel P and Anderson P W 1962 Phys.Rev. **125**, 1263-71.

Mott N F 1973 in 'Cooperative Phenomena' dedicated to Herbert Fröhlich, ed. Haken H and Wagner M, Springer-Verlag, 2-14;
1987 Nature **327**, 185-86;
1990 Adv.Phys. **39**, 55-81;
1991 Phil.Mag. Lett. **64**, 211-19;
1993a Physica C**205**, 191-205;
1993b Phil.Mag.Lett. **68**, 245-46.

Mott N F and Davis E A 1979 Electronic processes in non-crystalline materials, 2nd edn. Oxford University Press, Oxford.

Mott N F, Davis E A and Street R A 1975 Phil.Mag. **32**, 961-96.

Mott N F and Gurney R W 1940 Electronic Processes in Ionic Crystals, Oxford: Clarendon Press.

Mustre de Leon J, Conradson S D, Batistic I and Bishop A R, 1990 Phys.Rev.Lett. **65**, 1675-78.

Mustre de Leon J, Batistic I, Bishop A R, Conradson S D and Trugman S A 1992 Phys.Rev.Lett. **68**, 3236-39.

Nambu Y 1960 Phys.Rev. **117**, 648-63.

Nasu K 1985 J.Phys.Soc.Japan **54**, 1933-43.

Newns D M, Tsuei C C, Pattnaik P C. and Kane C L 1992 Comments Cond.Mat.Phys. **15**, 273-302.

Ogg R A Jr 1946 Phys.Rev. **69**, 243-44.

Olson G G *et al.* 1990 Phys.Rev. B**42**, 381.

Osofsky M S *et al.* 1993 Phys.Rev.Lett. **71**, 2315-18; 1994 ibid **72**, 3292.

Overend N, Howson M A and Lawrie I D 1994 Phys.Rev.Lett. **72**, 3238-41.

Pashitskii E A 1968 Zh.Eksp.Teor.Fiz. **55**, 2387-94 (1969 Sov.Phys. JETP **28**, 1267-71).

Peierls R E 1933 Z.Phys. **80**, 763-91.

Pekar S I 1951 (Russ.original Gostekhizdat), Research in Electron Theory of Crystals, US AEC Report AEC-tr- 5575 (1963)

Pines D 1961 The Many-Body Problem, Benjamin/Cummings, Reading, Massachusetts.

Pippard A B 1953 Proc.Roy.Soc. A**216**, 547.

Popov V N 1976 Kontinualnie Integrali v Kvantovoi Teorii Polia i Statisticheskoi Fizike, Atomizdat, Moscow, p.159 (in Russian).

Prelovsek P, Rice T M, and Zhang F C 1987 J.Phys. C**20**, L229-

33.

Radtke R J, Levin K, Schüttler H-B and Norman M R 1993 Phys.Rev. **B48**, 15957-65.

Ranninger J and Thibblin U 1992 Phys.Rev. **B45**, 773.

Rashba E I 1957 Opt.Spectr. **2**, 75.

Ray D K 1987 Phil.Mag.Lett. **55**, 251-56.

Robaszkiewicz S, Micnas R and Chao K A 1981 Phys.Rev. **B23**, 1447-58; ibid **24**, 1579-82.

Rossat-Mignod J, Regnault L P, Bourges P, Vettier C, Burlet P and Henry J Y 1992 Physica Scripta **45**, 74-80.

Ruscher C, Salje E, and Hussain A 1988 J.Phys. C**21** 3737-49.

Salamon M B, Inderhees S E, Rice J P and Ginsberg D M 1990 Physica A**168**, 283.

Salje E and Güttler B 1984 Phil.Mag. B**50**, 607-20.

Sawatzky G A 1989 Nature **342**, 480-83.

Scalapino D J 1969 in Superconductivity **1**, ed. Parks R D, Marcel Deccer, 449-560.

Schafroth M R 1955 Phys.Rev. **100**, 463-75

Schlesinger Z, Collins R T, Rotter L D, Holtzberg F, Welp U, Grabtree G W., Liu J Z, Fang Y and Vandervoort K G 1991 Physica C**185 − 189**, 57-64.

Schmitt-Rink S, Varma C M and Ruckenstein A E 1988 Phys.Rev.Lett. **60**, 2793-96.

Schnelle W, Braun E, Broicher H, Dömel R, Ruppel S, Braunisch W, Harnischmacher J and Wohlleben D 1990 Physica C**168**, 465-74.

Schrieffer J R 1964 Theory of Superconductivity, Benjamin, New York.

Schrieffer J R, Wen X G, Zang S C 1989 Phys.Rev. **B39**, 11663-79.

Sewell G L 1958 Phil.Mag. **3**, 1361-80.

Shoenberg D 1952 Superconductivity, Cambridge University Press, 78-86.

Street R A and Mott N F 1975 Phys.Rev.Lett. **35**, 1293-96.

Sugai S 1991 Physica C**185 − 189**, 76-79.

Svensson E C 1984 Proceedings of the 1984 Workshop on High-Energy Excitations in Condensed Matter, Los Alamos National Laboratory publication LA-10227-C, **11**, 456.

Taliani C, Pal A J, Ruani G, Zamboni R, Wei X and Vardini Z V 1990 Electronic Properties of HT_cSC and Related Compounds, Springer Series of Solid State Science (Springer-Verlag, Berlin) **99**, 280.

Tateno J 1993 Physica C**214**, 377-84.

Thomas G A, Rapkine D H, Cooper S L, Cheong S-W, Cooper A S, Schneemeyer L F, and Waszczak J V 1992 Phys.Rev. **B45**, 2474-79.

Tjablikov S V 1952 Zh.Eksp.Teor.Fiz. **23**, 381.

Toby B H, Egami T, Jorgensen J D, and Subramanian M A 1990 Phys.Rev. Lett. **64**, 2414-17.

Tolmachev V V 1958 in Bogoliubov N N, Tolmachev V V and Shirkov D V 'A New Method in the Theory of Superconductivity', Academy of Science (Moscow) (1959 Consultant Bereau, New York).

Toyozawa Y 1961 Prog.Theor.Phys. **26**, 29-44.

Uchida S 1992 J.Phys.Chem.Solids **53**, 1603-10.

Uemura Y J *et al.* 1991 Nature **352**, 605-607.

Varma C M, Littlewood P B, Schmitt-Rink S, Abrahams E and Ruckenstein A E 1989 Phys.Rev.Lett. **63**, 1996-99.

Verbist G, Peeters F M and Devreese J T 1991 Phys.Rev. **B43**, 2712-20.

Vigren P T 1973 J.Phys. **C6**, 967-75.

Von Molnar S, Briggs A, Floquet J, and Remenyi G 1983 Phys.Rev. Lett. **51**, 706-09.

Wheatley J M, Hsu T C and Anderson P W 1988 Phys.Rev. **B37**, 5897-900.

Wood R F and Cooke J F 1992 Phys.Rev. **B45**, 5585-606.

Xiang X-D, Vareka W A, Zettl A, Corkill J L, Cohen M L, Kijima N and Gronsky R 1992 Phys.Rev.Lett. **68**, 530-33.

Xing D Y and Liu M 1991 Phys.Rev. **B43**, 3744-47.

Yamashita J and Kurosawa T 1958 J.Phys.Chem.Solids **5**, 34.

Yu R C, Naughton M J, Yan X, Chaikin P M, Holtzberg F, Green R L, Stuart J and Davies P 1988 Phys.Rev. **B37**, 7963-65.

Yu R C, Salamon M B, Lu J P and Lee W C 1992 Phys.Rev.Lett. **69**, 1431-34.

Zhang X and Catlow C R A 1991 J.Mater.Chem. **1**, 233-38.

Zhang F C and Rice T M 1988 Phys.Rev. **B37**, 3759-61.

Zheng H, Avignon M and Bennemann K H 1994 Phys.Rev. **B49**, 9763-73.

Zheng H, Feinberg D and Avignon M 1989 Phys.Rev. **B39**, 9405-22;
1990 ibid **41**, 11557-63.

Author index

Subject index

Printed and bound by CPI Group (UK) Ltd, Croydon, CR0 4YY

18/10/2024

01776256-0020